PRAISE FOR
THE HUNDRED-YEAR WALK

"In her remarkable book, *The Hundred-Year Walk*, Dawn MacKeen has taken the Armenian genocide and shown us its terrifying flesh, blood, bone, and sinew. Her vehicle is her grandfather's forced deportation, and she uses it to take the reader on a horrific ride into the heart of one of history's darkest moments."
— S. C. Gwynne,
author of *Empire of the Summer Moon*

"By telling the riveting story of her grandfather Stepan... Dawn MacKeen drives home that we're all part of the human family. *The Hundred-Year Walk* is an unforgettable contribution to the literature of suffering and memory."
— David Talbot,
author of *The Devil's Chessboard*

"A haunting journal of remembrance... MacKeen doesn't shirk from recounting the grisly details of genocide, describing brutal beatings, hunger to the point of cannibalism, and thirst to the point of urine-drinking. With a health-care reporter's deft touch, she manages to play down the utter pathos, but her dedication to baring gruesome facts is as unfailing as her loyalty to the mission thrust upon her."
— *Barron's*

"MacKeen weaves multiple historical sources for corroboration and context, but her main material, Stepan's unpublished memoir, lands the emotional punch of personal narrative. MacKeen's added perspective is what makes this book though. A moving portrait of one family's relationship to the past that offers surprising hope for reconciliation."
— *Globe and Mail*

"Gripping."   — *Outside*, a "Can't-Miss" book

"The highs and lows in her book are thrilling, as they compare and contrast the life of a man in the Ottoman Empire struggling from one moment to another in order to survive, as well as his American grand-daughter who makes this journey, albeit with as many modern comforts as possible. The linking of the two is the centerpiece of the book as MacKeen is able to viscerally experience the memories imparted from her grandfather."   — *Armenian Mirror-Spectator*

"This previously untold story of survival and personal fortitude is on par with Laura Hillenbrand's *Unbroken*. Further, this is a tale of tracing your family roots and learning about who you are. It will have broad appeal for a wide range of readers."   — *Library Journal*, starred review

"Readers will find themselves drawn into the whirlpool of events, soon forgetting the author's presence . . . powerful, terrible stories about what people are willing to do to other people — but leavened with hope and, ultimately, forgiveness."   — *Kirkus Reviews*

THE HUNDRED-YEAR WALK

An Armenian Odyssey

DAWN ANAHID MacKEEN

Mariner Books
An Imprint of HarperCollins*Publishers*
Boston   New York

www.marinerbooks.com

Mariner Books
An Imprint of HarperCollins Publishers, registered in the United States of America
First Mariner Books edition 2017
and/or other jurisdictions.

This is a work of nonfiction. All dialogue, details, and events
in the life of Stepan Miskjian are directly culled from his
memoirs, interviews, or other historical works.

Note: Some names and locations have been changed
to protect sources in Syria.

Library of Congress Cataloging-in-Publication Data
MacKeen, Dawn Anahid.
The hundred-year walk : an Armenian odyssey / Dawn Anahid MacKeen.
pages cm
Includes bibliographical references.
ISBN 978-0-618-98266-0 (hardcover) — ISBN 978-0-544-58292-7 (ebook)
ISBN 978-0-544-81194-2 (pbk.)
1. Miskjian, Stepan, 1886–1974. 2. Armenian massacres, 1915–1923.
3. World War, 1914–1918 — Armenia. 4. Armenian massacres
survivors — Biography. 5. Young men — Armenia — History — Biography.
6. Escapes — Armenia — History — 20th century.
7. Desert survival — Syria — History — 20th century. 8. MacKeen,
Dawn Anahid — Family. 9. MacKeen, Dawn Anahid — Travel — Turkey.
10. MacKeen, Dawn Anahid — Travel — Syria. I. Title.
DS195.5.M315 2016
956.6'20154092 — dc23
[B]
2015016713

Book design by Greta Sibley
Map by Mapping Specialists, Ltd.

Printed in the United States of America
22 23 24 25 26 LBC 8 7 6 5 4

To my mother, Anahid

CONTENTS

Constantinople

Adabazar
(The Miskjian's home)

Eskishehir

A N A T O L I A

Chai
(The Miskjian
family's camp)

Konya

Bozanti

TAURUS MOUNTAINS

Adana

Amanus Mountains

Aleppo

MEDITERRANEAN
SEA

Stepan's Wartime Route

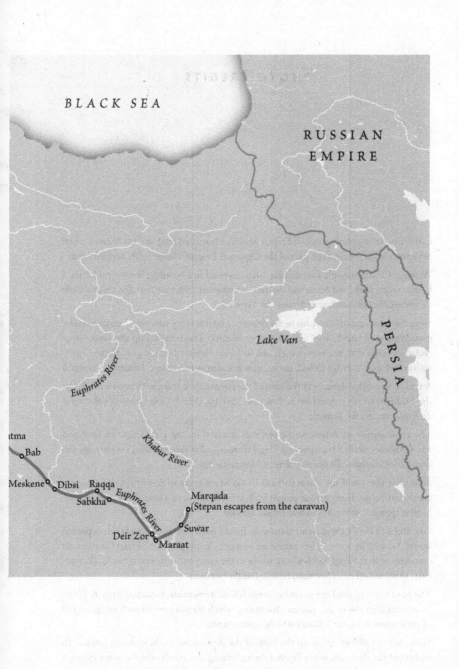

PHOTO CREDITS

Earliest surviving photograph of Stepan Miskjian (pictured left) and his friends, taken circa 1910 in Adabazar, then part of the Ottoman Empire. *Courtesy of the Miskjian Family*

When in Constantinople for business, Stepan stayed in a boarding house on Mahmud Pasha Street, in the old Stamboul district. *Armenians in Turkey 100 Years Ago, Orlando Carlo Calumeno Collection & Archives & Birzamanlar Yayıncılık*

Stepan would regularly carry his heavy deliveries up this steep street in Constantinople, past Armenian, Greek, and Turkish shops. In 2007, Dawn huffed up the same street and discovered that, not surprisingly, she had less stamina than her grandfather. *Armenians in Turkey 100 Years Ago, Orlando Carlo Calumeno Collection & Archives & Birzamanlar Yayıncılık*

A typical street in Adabazar at the turn of the twentieth century, with homes similar to the Miskjian family's. *Armenians in Turkey 100 Years Ago, Orlando Carlo Calumeno Collection & Archives & Birzamanlar Yayıncılık*

The train station in Adabazar. Stepan was arrested by the police here in 1913 and wrongfully accused of transporting illegal weapons. *Armenians in Turkey 100 Years Ago, Orlando Carlo Calumeno Collection & Archives & Birzamanlar Yayıncılık*

Soon after the round-up of intellectuals in Constantinople in April 1915, the mass deportations began. Here, armed guards lead Armenians out of a town in eastern Anatolia that spring. *Courtesy of the Ararat-Eskijian Museum*

Like Stepan's family, these Armenians were forced from their homes and then deported across Anatolia by train. "They crammed 880 people into ten cars," Franz Günther, an executive with the Baghdad Railway, wrote to the company's chairman on October 30, 1915. *Deutsche Bank AG, Historisches Institut, Frankfurt am Main*

The hard labor of road construction often fell to Armenians, including Stepan. Here, Armenian men toil in the Taurus Mountains, which Stepan crossed on foot in the fall of 1915. *Armin T. Wegner © Wallstein Verlag, Göttingen 2011*

Women and children made up the bulk of the Armenians in these desert camps. To withstand the elements, many layered their clothing, and much of it was worn through shortly after the deportation began. *Armin T. Wegner © Wallstein Verlag, Göttingen 2011*

Marched to the Syrian desert, an Armenian mother mourns her dead child. *Library of Congress, Prints & Photographs Division, George Grantham Bain Collection, LC-USZ62-48100*

Human bones found by farmers outside Ras al-Ayn, another slaughterhouse north of Deir Zor, Syria. © *Bardig Kouyoumdjian. "Human bones put aside by the local farmers." From* Deir-Zor: On the Trail of the Armenian Genocide of 1915

Stepan's three sisters and nieces. Top row: Araxi, Mari, and Louise. Lower row: Arshaluys, Madeline, and Aghavni. This was taken around 1921, the year they left Constantinople and immigrated to the United States, laying the groundwork for Stepan and his mother to follow. *Courtesy of the Miskjian Family*

"God Bless America": Stepan and his family finally arrive in the United States; here, they stand near their home in Spanish Harlem in 1931. From left: Stepan, Anahid, Alice, and wife, Arshaluys. *Courtesy of the Miskjian Family*

Newly widowed, the 85-year-old Stepan stands beside his daughter, Anahid, and granddaughter, Dawn, in 1971. *Courtesy of the Miskjian Family*

A typical street in Chai, Turkey. Not far away from here — at least in terms of mileage — sprawled the internment camp the Miskjian family was deported to in 1915. Alongside thousands of others, they lived in makeshift tents beside the train station. *Courtesy of the author*

Down the street from the store in Adana, Turkey, where Dawn had her headscarf makeover. *Courtesy of the author*

The caretaker of the genocide memorial in Marqada, Syria, unlocks the door. The chapel was built beside the hill where Stepan's caravan was massacred in 1916. He escaped by crawling on his hands and knees and traveling across the desert, with almost no water, for the next six days. *Courtesy of the author*

The village in Syria that is home to the descendants of Sheikh Hammud al-Aekleh, who saved Stepan after the massacres and became like a father to him. Finding the family in 2007, Dawn had the opportunity to thank them for saving her grandfather's life. *Courtesy of the author*

Dawn and descendants of Sheikh Hammud al-Aekleh enter the cemetery in Syria, where the revered chieftain is buried. Also pictured: the Bedouin envoy from Raqqa who helped to locate al-Aekleh's family. *Courtesy of the author*

At the cemetery in Syria, Dawn pays her respects to a relative of Sheikh Hammud al-Aekleh in 2007. *Courtesy of the author*

On April 24, 2015, the hundredth anniversary of the Armenian genocide, Dawn and her mother, Anahid (pictured), marched alongside an estimated 160,000 others to raise awareness of the killings. It was one of the largest demonstrations in Los Angeles history. *Courtesy of the author*

A NOTE ON THE TRANSLATION

THIS BOOK SHIFTS between the early-twentieth-century Ottoman Empire and the present day. All languages in these pages have been transliterated and translated for ease of reading rather than technical precision. That includes my grandfather's primary languages, western Armenian and Ottoman Turkish. In most cases, the Armenian ու has been transliterated as a *u*; thus the *u* in the name Harutiun is pronounced like the French *ou*. On the other hand, the name of the book's main character, my grandfather Stepan Miskjian, is pronounced *Ste-pan Meesk-geean*. Modern Turkish, the language spoken in Turkey today, uses the Latin alphabet, unlike Ottoman Turkish. It, too, has been transliterated for easier pronunciation; for example, the name Cemal has been changed to Jemal, and my grandfather's town, Adapazarı, has been changed to Adapazaraï, the last syllable of which is pronounced like the *a* in *serial*. Ottoman endings in *d* and *b*, which often appear as *t* and *p*, respectively, in modern Turkish, have been left unmodified in the chapters that take place during Ottoman times. The Arabic has been transliterated to the most common forms. In addition, all dates have been changed, as best as possible, from the Julian calendar, used during my grandfather's time, to the Gregorian calendar.

PART ONE

Before

The Lost World

2006

FOR AS LONG AS I CAN REMEMBER, my mother has been talking to her dead parents. Growing up, I would find her in the kitchen, locked in conversation with Mama and Baba. At the sink, her hands scrubbing a dish, her voice a murmur. So it was no surprise when, in the summer of 2006, I stumbled on her again like this. It had been just a few weeks since I had moved back into my childhood home, and there I was in the doorway trying to eavesdrop, just like I had back in grade school. Only now I was thirty-five. I couldn't quite make out her words, drowned as they were by running water and the clank of Corelle plates. Oblivious to me standing there, my mother continued to shake her cropped brown bob back and forth, moving her lips furtively.

"*Inch ge medadzes,*" she said, shaking her head, the Armenian words sounding like gibberish to me.

"Are you talking to them again?" I asked.

"Yes," she replied, her mood perennially upbeat. "I ask them for advice, and they always give it to me. They are my spirit guides, Dawn. They should be yours too!"

I rolled my eyes and we both laughed, not taking ourselves too seriously. In the weeks since I'd left my bustling life in New York and returned to the Los Angeles house where I had been raised, my mother's otherworldly talks had become part of my universe again. I'd forgotten the never-ending surprises of life with my small but plucky mother, Anahid. Spontaneous and excitable, she could transform a drab doctor's office or a corner diner into a party, just by raising her arms and breaking into dance.

My father, Jim, and I would remark that she was the last person you'd expect to be a probation officer. She was unflinchingly positive about the human capacity for goodness, allowing the petty criminals she supervised to get away with nearly anything on her watch. She'd devoted her life to helping people. Not only her clients, but also Armenian immigrants unfamiliar with the customs of the United States. Our phone was constantly ringing. She'd taught my American father and me just enough of the language for us to say "One moment" in Armenian — *Meg vayrgean* — when people called and started prattling away about needing a ride to the doctor, the lawyer, or the green-card office.

Despite the comfort of being back in my roomy, Spanish-style home, the initial excitement had worn off. Huddled under my flower-print bedspread, surrounded by high-school soccer trophies and my homecoming-princess tiara, I felt like a character in a dark comedy about an aging prom queen who returns to her childhood home after flaming out in the big city. By the hour, my life in New York felt farther away — my morning runs through snowy Central Park before work; my deadline hustle to file yet another health-care story at my magazine job; my race to meet friends after work for a wine-fueled late dinner somewhere dark and candlelit. For years, my life in New York had felt like a sprint in a marathon that I never wanted to stop. It was what I craved; it was what I thought I needed; it was why I'd left my home and moved across the country in the first place.

But shortly after my birthday the previous February, something had changed. I'd never paid much attention to my mother's calls to come home, but suddenly I couldn't ignore her anymore. Perhaps it was her advanced age (she was then seventy-eight). Or maybe it was my own realization that, as a reporter, I was spending my life telling other people's stories and ignoring my own family's incredible one.

Because my grandfather had died when I was a toddler, what I knew about him was mostly family legend. Countless times, I had heard the dramatic tales from my mother of how her father, Stepan Miskjian, had wandered in the desert of what is now Syria, how he had staggered across it for a week on nothing but two cups of water. How he had led a group of Armenians to safety, away from the Turks who wanted to kill them.

She'd repeat this tale on loop. As she saw it, any occasion — during a morning bowl of Cheerios or after a piece of birthday cake — was the right time to recount her father's near-death experience.

His story had truly haunted her childhood too, when days would begin and end with Baba in tears as he retold what he'd witnessed. He made a new home for his family in Spanish Harlem, but they were so poor she slept in a hammock. Perhaps looking into his daughter's innocent face reminded him of the thousands of children in their orphan uniforms who had paraded past him in the camps on their way to be slaughtered. He had lost almost everything in the ethnic cleansing; all he had was his story. This was our family's heirloom, our most precious bequest, and it was inherited by every subsequent generation — along with the burden of telling it again.

Still, as a kid, I retained nothing from the much-repeated saga but the single detail that he'd drunk his own urine to survive in the desert. Repulsed, I'd always ask, "Why would anyone do that?"

"It's because he was Armenian and faced very difficult times," my mother would explain. "It's all here."

And then she'd pull out two small booklets published by an Armenian press in the 1960s: her dad's firsthand account of his survival, focused on the period when he was fleeing the Turks in Mesopotamia. I would stare at the hundreds of pages of Indo-European script, unable to cross the language barrier and uncover the secrets of his memoir, a narrative he'd begun writing in the 1930s and continued working on for the rest of his life.

My mother had spent many years attempting to translate these booklets into English. This wasn't just her personal desire to share our family's trials but part of an attempt to educate the world and ensure that ethnic cleansing never happened again. Her father's story was the story of the forgotten genocide. The trains stuffed with people, the death marches, the internment camps. All were familiar horrors to me, to my generation, but the images I'd seen were from the Holocaust of World War II. As the Jews would be, the Armenian minority had been demonized as a threat to society. With the global tumult of World War I used as a cover, the plan was set in place. The majority of the two million Ottoman Armenians had been forced from their homes and deported to barren regions they had seen only on maps, if at all.

From 1915 to 1918, an estimated 1,200,000 Armenians perished. Those who managed to stay alive were scattered across the globe. My mother's surviving aunts and uncles lived in Turkey, France, and the United States — something I had previously thought was a little glamorous. After learning more about my family history, I found it heartbreaking. Entire families had been lost or severed from one another. Stateless, some of them drifting like embers after a fire, the rest just ashes. Adolf Hitler, before his invasion of Poland in September 1939, said: "Kill without pity or mercy. Who still talks nowadays of the extermination of the Armenians?"

In a way, der Führer was right. Only the Armenians seem to remember the Armenians.

"Help me, Dawn. Help me tell Baba's story," my mother would often say as the years went by and I embarked on a career as a reporter.

"Please help me," she repeated on the phone when I was living in Chicago, San Francisco, DC, New York. "Almost done, Dawn, we're going to print this," she informed me when I came home every Christmas. In truth, despite her determination, her translation would never be finished; her father's language was too difficult. She could speak Armenian fluently but she could barely read the characters; she had learned the language from her mother but had never formally studied it. However, that didn't stop her from trying. She grew ever more resolute as Turkey's denials of genocide grew in volume, with its claims that the Armenians had staged an uprising and sided with the Russian enemy during a time of war and that their deaths were an unfortunate consequence of that. She watched as her parents and her parents' friends and nearly all the other eyewitnesses died.

To my mother and many Armenians, *genocide* is an important word. *Massacres, atrocities, killings* — all these words fail to describe "the murder of a nation," as Henry Morgenthau, the American ambassador to Constantinople, called it at the time. Nearly twenty countries, including France and Canada, officially used the term, which was created during World War II by a Polish Jew, Raphael Lemkin, to describe the extermination of the Armenians and the Jews.

I didn't understand my mother's commitment, her sense of urgency, until April 2004, when a distant relative finally intervened and translated my grandfather's two booklets. I'll never forget the cross-country flight when I read what my grandfather Stepan remembered, in his own words. His stories transported me back to a very different era, to the last days of the Ottoman Empire, when the Great War consumed much of the world. "Being a witness to that satanic pogrom, I vowed it as my duty to put to paper what I saw," my grandfather wrote in the introduction to his memoirs in a straightforward, unemotional style that I would come to know well. With his extraordinary memory, he described his road to hell, relating his conversations with the guards overseeing the death march and his dying friends' last utterances. As I turned the pages, I felt sickened by what he had endured. I slowly came to understand what it meant: my mother existed, I existed, my aunt, my uncle, my cousins, my cousins' children — my entire family existed because he had survived what so many had not.

Nearly a century later, where was my sense of moral obligation? Doing nothing felt like forgetting, and forgetting genocide seemed almost as heinous as the crime itself, especially in light of Turkey's denials. I read these booklets, which represented only a portion of his remarkable life, and although I did not move back to Los Angeles for another two years, the seeds were sown. I had to help my mother tell Baba's whole story. I had to better understand the past.

I finally moved across the country in May of 2006, and soon enough, that summer seemingly evaporated in a haze of Los Angeles heat, suffused with my mother's supernatural talks at the kitchen sink. By August, I was deep into my routine of squandering days at a café sipping double nonfat lattes and reading about the genocide. My plan upon arriving home had been to fill in the gaps in my grandfather's stories and begin shaping them into a proper account. But the plan was quickly stymied by a near total lack of information. Most of the witnesses and survivors were dead; their children too. The bulk of the primary accounts were written in a language I didn't understand. And though Baba's memoir was powerful, it recorded just a few years of his life and ignored other critical periods of the ethnic cleansing. There were so many missing pieces in his story, gaps that left us wondering what had transpired before he landed in an internment camp in the middle of the war. Eager to help, my mother had been calling all her friends, trying to locate survivors from her parents' hometown for me to interview.

It was an almost hopeless task, and the only lead she had was one old lady she remembered who had known my grandfather and could possibly tell us more about the deportations from their hometown.

One afternoon, following my daily coffee binge, I returned home depressed by the reading material and frustrated by my lack of progress. In the living room, near the stained-glass windows, my father, a retired auditor, was leaning back in his leather La-Z-Boy, feet up, *Consumer Reports* in hand. From where I stood, only the crown of his head was visible. Across the room, my mother sat on the lime-green sofa turning the pages of the *Los Angeles Times*. The green shag carpet I remembered from my childhood had long ago been ripped out, but almost everything else seemed frozen in time. My mother could never throw anything away; even the smallest items had value. She rarely bought new clothes or furniture, an austerity measure that I was beginning to understand was a symptom of her having had nothing as a child in Istanbul and making do with hand-me-downs.

That afternoon, I walked off to my room without saying a word to the good people who'd raised me. The house was hot — of course it was — and that just stoked my irritability. Flinging myself onto my bed,

I stared at the pictures on the wall of my wrinkle-free younger self with my best friends from the neighborhood.

I felt like I was regressing. Shortly after arriving, I had started acting like a paranoid teenager, speaking on the phone in a whisper so my parents could not hear, listening for the creak in the hallway that told of intergenerational eavesdropping. Whenever I went out, I gave them my itinerary and the hour of my expected return, as I had in high school. *Just a little longer,* I'd tell myself. *Only until I can get on my feet and rent my own place.*

I blamed my mother for my sudden mess of an adult life, the way one irrationally faults one's parent for everything. After all, it was her insistent devotion to her father's history that had motivated me to come home.

That Thursday, as I lay there stewing, I was seized by a horrible thought: *Did I make a mistake moving back?* I suspected I had; I was jealous of my friends' rising careers and busy social lives. Pushing it out of my mind, I moped back into the living room.

"How was your day, Dawn?" my mother asked lovingly. "Do you want something to drink?"

"No."

"Something to eat?"

"No."

"Oh, Dawn." She followed me to the long dining room table and sat down in the hard ornate wood chair that she had inherited from her parents more than thirty years earlier.

"I called the Ararat nursing home today, and that old woman we thought might help us . . . she has died."

I frowned at the news.

"She was about a hundred and two years old," my mother added. "Don't worry, honey. You can find someone else to interview," she said before I had a chance to respond.

"Who do you suggest?" I said flatly. "Your father is dead. His siblings are dead. Now the lady in the nursing home is dead. They're all dead."

She didn't say anything.

"Mom, I can't help you with Baba's book."

"Why not, Dawn?"

"Because everyone is dead!" That was it, I told myself. I had put my real life on hold for long enough. I had to move on and start looking for a proper job.

"It's okay, we'll find someone else. Don't give up. You can do it," she said.

We continued this cycle, repeating the same things, raising our voices — mine always louder than hers.

"I can't help you!" I began to cry. Then a voice welled up deep inside of me. It burst out as if I weren't the one choosing the words: "I cannot help you unless you raise your father from the dead and have him tell me what exactly happened to him."

Silence. What could she say? Dropping her gaze to the hardwood floor, she finally begged, "Please, Dawn. Please," as the shame of my outburst settled on me.

Two days later, my mother woke early and strode to the bookshelf in the dining room where my parents kept their lifetime collection of novels, alternative-medicine encyclopedias, and how-to manuals. Row by row, she began to take them all out, inspecting each one and then throwing it to the ground, not completely knowing what she was look-ing for, knowing only that she had to look. She rose up on her toes and reached into a corner of the bookcase, behind some pamphlets. There, to her surprise, she discovered two small notebooks, never be-fore opened by her, bearing her father's writing.

"Look what your mother found," my father said later that Saturday mor-ning. His blue eyes were gleaming behind his bourbon-colored glasses as he stood outside my door, clearly waiting for me to emerge from my room. I didn't answer, still angry and embarrassed by the other day's tantrum. I remained quiet when I found my mother at the dining-room table, books strewn everywhere, as if unshelved by an earthquake.

"It's my father's writing," my mother said under her breath about the notebook in her hand. It was the first time we had spoken in the forty-eight hours since our blowup. In all my years, I've never seen her so subdued. Subdued just isn't something she does.

"What do you mean, Mom?"

She slid one of the books, speckled in black and white like confetti, over to me.

"These are his notebooks. I must have put them there after he died, not knowing what they were."

I opened it up. Black cursive script in Armenian, precise penmanship and clean lines of text. Not being able to actually read the text, I focused on the aesthetic details of his script. I'd previously assumed he was like her, abstract, carefree, since the booklets I'd previously seen had been typeset, devoid of his penstrokes. I must have said something aloud, because my mother responded.

"Yes, Dawn, Baba was very neat."

On the very last page, there were titles and page numbers.

"This looks like a table of contents," I said.

My mother studied it. "Yes, that's what I thought too."

He'd numbered the pages and printed the years at the top of each one; 1910 was on page 15; 1911, page 37. These were memories of his early years that he'd recorded in the genocide's aftermath. It was as if he'd been trying to reconstruct his old life, chapter by chapter.

My mother held the two notebooks with both hands. They were so thin. *So easy for them to disappear,* I thought. After all these years — almost my entire lifetime — they had reappeared at this very moment when I needed them most.

"I can't believe this. I just can't believe this," I repeated. "Mom, what do you think it's about?"

She flipped through the pages again and struggled to read her father's writing, her voice barely above a whisper. "*Ashkharhen e . . .*' 'The world is . . .'"

In the coming days, my mother and I scrambled to the homes of her friends and relatives in the close-knit Armenian community, the newly discovered notebooks in hand. This quickly became my social scene in LA: everyone I hung out with was upwards of seventy years old, fluent in Armenian, and willing to translate upon request. Doilies and plastic tablecloths were often present, and all my new best friends had an insatiable curiosity about my love life. Still single at thirty-five? Why had I not found a man? They knew just the right person! This refrain echoed into a full chorus. My mother and I kicked off each bright morning with Arlene, a petite woman with perfectly coiffed dyed-black hair who lived two blocks away. After meeting us at the door with a warm kiss, she ushered us into the elegant dining room. As we sat down, she disappeared into the kitchen, then reemerged with little cups of Armenian coffee and sweets on a tray. I spooned a touch of sugar into the coffee, took a sip, and then set the cup down on the saucer. During the first visit, the endearing Arlene immediately launched into a description of her two nephews — both available! She was so kind and wanted only to help, but as we sat there, I started thinking about an old Armenian tradition that my mother, always the comedian, had threatened me with. "If you don't watch it, we are going to arrange a *khosg gab* for you!" She could barely finish without cracking herself up. "No, no," I'd say, feigning fear. A *khosg gab*, as she explained, was basically an arranged engagement. The girl's answer to the proposal could be found in how she prepared her suitor's cup of coffee. Did she add sugar or not? If the man tasted sweetness, she liked him, and wedding bells soon rang. My own mother had forgone the tradition, but my parents did marry in the Armenian Apostolic church. Since the ceremony was in Armenian, my poor father had no idea what he was agreeing to; he didn't realize it was time to say "I do" until he received a good jab in the ribs from my mother. Over the years, the tall, lean man of hearty Scottish-English descent grew to love my mother's culture and its name for non-Armenians — *odar* — which, he joked, sounded suspiciously like *odor*.

Once she opened up the notebooks, Arlene translated my grandfather's writing with ease. Arlene read chapters with titles like "How I Became a Courier," "An Anecdote from Pera, Constantinople," and "Return from the Armash Market." This last one was about how Stepan and his good friend Nerses Aghajanian did not want to pay for a second cart back home after a trip to the market with his family. To fit all their goods — and relatives — into one cart, they stuck Nerses's parents inside two barrels. The story unfolded like a cartoon, of course:

while the cart was going up a hill, the precious human cargo rolled off and tumbled down, the dad yelping all the way, though both of them were, surprisingly, uninjured.

How different this time in his life had been, I thought. *Before what is to come.*

So far, the anecdotes Arlene read to us were all from the years leading up to war in Baba's hometown, then called Adabazar, but now Adapazarï. He recounted each moment almost like a journalist covering an event — the time of day, the number of people in attendance, the full names of friends and acquaintances. Later on, as I grew more familiar with his story and its horrific conclusion, I became obsessed with learning what had happened to these people, even placing advertisements in national and local Armenian newspapers to track down their descendants.

It was the beginning of my own quest, one that would last almost a decade and would take me to a library in Paris; to the Euphrates River in Syria, where I was followed by the secret police; to a snowy Viennese monastery populated by compassionate Armenian monks, where I was so cold my fingers went numb as I thumbed through a card catalog in their library.

We also relied on my mother's cousin Yevkine to help with the translation, as we wanted to spread out the time-consuming burden. She lived high up in the Santa Monica Mountains, and we often gathered in her sunlit breakfast nook, the brightest room of the house, to pore over the notebooks. At first, Yevkine struggled with the script; she hadn't read Armenian since she was a teen in Beirut. But then it came flooding back. Faster and faster, she translated my grandfather's words into English. Occasionally she broke into hysterics, shaking from laughter, her eyes tearing up. "What?" I'd ask. "What?" My mother was giggling too, because Yevkine read the Armenian aloud first. "Tell me," I pleaded.

"Your grandfather told a priest that his friend was a deacon — when he wasn't. In the middle of the service, they called him up, and the man couldn't lead the hymn."

"Mom, he was a little prankster!" I said. I couldn't believe it. All I knew of him was his struggle, his pain. Somehow, I had reduced him to one dimension: he was a survivor. I hadn't thought of him — or anyone else who endured a genocide — as having a personality, as being funny and knocking back stiff drinks with pals. That's what a holocaust does — it erases.

Sometimes, as Yevkine was reading aloud, she paused and looked upward, as if the sky held the right word. I kept transcribing, always asking questions. Occasionally, I would catch a word. This was a feat, since my fluency in Armenian remains confined to the crucial words a toddler might use, such as *shun* (dog), *gadu* (cat), *got* (milk), *vorig* (butt), and *vardig* (underwear). During one particular meeting, my mother suddenly looked at me and then interrupted Yevkine as if she had made an important discovery: "Baba was very meticulous and exact. You're just like him, Dawn, in your exactness in telling his story," she said, her eyes on me. "You have that same drive to be totally accurate. I just thought of that. Baba was very organized. You're the same, Dawn."

She was doing it again, encouraging me, connecting me to the past. However, this time I didn't dismiss her. Something had occurred when we discovered those handwritten notebooks that I cannot — and will never be able to — fully explain. Somehow, it felt like my grandfather was with us, leading the way, alive. As Yevkine pressed onward to another section, my mother continued to confer with her dead father, quietly looking off to the side, conversing under her breath, until she finally declared, "Baba approves of what you are doing."

I was only half listening to her; I was used to her otherworldly pronouncements, and my mind drifted elsewhere. It was something about the way my mother had said, "Baba was very meticulous and exact."

Then it hit me.

The two handwritten notebooks we had just found on my mother's bookshelves detailed his life up to the outbreak of the Great War, in 1914. The other booklets, the ones published by the Armenian press that my mother had spent years trying to translate, began in 1916 and went to the conflict's end. *A man so meticulous would not have left out two critical years,* I thought, trying to convince myself that there had to be more, somewhere. I desperately wanted to fill in this story's gap. "Mom, there has to be more of his writing. Can you search your house from top to bottom? How about Uncle Johnny?" He was her younger brother, and he lived a few miles away. "Can you ask him too?"

A week later, my uncle Johnny upturned his garage, and at the bottom of a box stuffed with his deceased father's belongings lay two more of the notebooks.

I don't know what I had expected. After all these years stashed away, one notebook remained remarkably intact; the other did not. It was yellowed and stiff, the edges curled up as if someone had held a flame to it.

Hopeful, though, we returned to see Yevkine. She knew how important this was to us, and generously spent days with us reading from the same cursive Armenian ink, precise and uniform. We often stayed there until large shadows fell around us and lights began to turn on one by one in the homes on the hillside. In the weeks and months that followed, this was all we did. Yevkine had known my grandfather and felt as pained as we did at his revelations. Often I could tell a terrible passage was coming just by glancing at her grimaced face. "'Like pickled sardines, we were lying on top of each other, without any sleep,'" Yevkine read to us, her expression sorrowful. "'The next morning, they took us out and tightened our bonds.'"

These last two notebooks, the ones from my uncle Johnny's garage, covered the missing time, from 1914 to 1915. Though the writer was clearly the same, his tone had turned somber since the war's outbreak. While his early words had related all his mischievous antics with friends, now he told of a darker time, alone. I could see the man he was becoming, the seeds of sadness planted.

For all my mother's devotion to Baba's story, these were details that she had never heard. Listening to the recounting of her father's awful days, my mother would stare somewhere far off, her narrow eyebrows frozen in an extended arch as we heard about how his feet were beaten with wooden boards. *Did he cry out?* I wondered. *For his mother? For God?* In other passages, we learned about how he was pushed to the river's edge to be shot by two gendarmes. We also listened to him describe the richest girls of his town and how they had carried the embroidered fabrics of their dowries from camp to camp until they realized there would be no future, no wedding, and had traded them for bread. I bit my lip and inhaled deeply to keep from crying.

Often, when Yevkine turned a page of the fragile notebook, the corner would snap off and fall softly to the sunburst-colored placemat underneath. Periodically, we would stop to tape the corners back onto the pages. At times, full sheets came away from their decades-old binding. Yevkine patiently proceeded despite holes in the paper that had swallowed entire words. She was as entranced as we were, having no option but to try to follow a road that was quickly being washed away.

"I had no idea," my mother would say, shaking her head. "I had no idea." We were both getting to know this side of my grandfather, two generations trying to piece his life together. It was as if my grandfather had understood — what he couldn't finish in his lifetime would be finished in the next.

Sometimes, we had a pile of the notebook's pieces and were unsure where to place them. We'd search for the matching triangular shape on the page or the interrupted sentence. "It's like a puzzle," one of us would say, each snippet — whether it was there or missing — revealing the shape of my grandfather's life and transporting me back to the past, where I increasingly found myself living. I wanted to leave the present, often cutting short evenings with my friends to go home and study the era. In addition to his account, I was consuming every memoir and history book I could find about the Armenian experience during World War I and those years from 1914 to 1918. Slowly, I began to see his story in four parts: His dreams for himself before the war. His subsequent conscription into a labor battalion, and the exile of his family from their home. His struggle to stay alive in the Mesopotamian internment camps. And the refuge that he found, so far from home, with an Arab sheikh who transcended their differences of religion and culture and welcomed him into his clan. With my grandfather's words as my guide, my day-to-day life receded into his. In my dreams, in my waking hours, it was a lost world that I began to inhabit.

He had written his entire story. Somehow, I just knew. As I sank deeper into his narrative, I could feel the dry air; I could see the earth closing in around me. I could feel my thirst, his thirst. I had to visit this place. I wanted to touch the land my grandfather had walked, drink from the Euphrates River like he had, because I had some sense that it would bring me closer to him. I also wanted to see what his life had been like before it was all stolen. I wanted to visit his tree-lined hometown where he had played tricks on his friends, the place that had shaped him. I wanted to see the green hills surrounding his Adabazar, where his dreams of becoming the town's first courier had taken root, before they withered and died in the desert. I was afraid, but I had no choice. My grandfather had left a road map to his life — all I had to do was follow it.

Empty Plans

1910–1912

THE LOCOMOTIVE RELEASED A SWIRL OF STEAM and edged down the tracks. In his seat, Stepan Miskjian settled in for the five-hour ride to Constantinople. The twenty-three-year-old Adabazar native took this journey frequently and knew it well — the long lake of Sapanja, the dense trees of the mountaintop, the brilliant blue of the Marmara Sea. As the train whipped past the countryside that March day, the magnificence of Anatolia flickered past him, illusory as a daydream.

In the distance, a cluster of red-tiled rooftops sloped down a hill, and he could see the station in the next town of Izmid teeming with passengers. Shortly after the train pulled up, the doors of the cars flew open, and a crowd bustled in and spilled into the remaining seats. From the platform, the travelers asked the heavily mustached Stepan about room in his compartment. "It's full," Stepan replied, though four spaces around him were empty. As the people continued to shout, the conductors escorted some aboard. Stepan tensed. He wanted to be able to stretch out but realized it would be impossible to keep the seats vacant. Better to share with fellow Armenians, he thought.

He leaned toward the window and called out in his native tongue to the swell of waiting Turks and Armenians. Understanding his words, three passengers quickly boarded and slid into the spaces beside him, and the train departed. The throaty churn of their language soon filled the car, and the conversation bounced around, inevitably landing on the men's livelihoods. The soft-spoken Stepan explained that he was a *perezag*, a peddler, on his way to buy skirts and blouses. The newcomers, in turn, relayed that they were *emanetjis*.

Inch e? What? Stepan's brown eyes looked flummoxed. He'd heard the word but was unsure of its meaning.

"We bring in goods to the merchants, transport parcels, deliver currency, whatever the merchants want," they said. "We either deliver things or pick them up."

Basically couriers, he understood, only they were depositaries too. "How are you able to transport so many pieces and still make a profit?" he asked. "The railway charges so much for luggage."

"We usually take the steamboat and don't pay for parcels, just five *ghurush* per passenger. But today we had a rush order and were obliged to take the train."

"How many of you do this type of business?" Stepan asked.

"There are four, and all of us make a living."

Characteristically, Stepan did the math. If the small town of Izmid could support four *emanetjis*, what about Adabazar? His town was twice the size of theirs. Already, he regularly traveled his *sanjak*, or county, selling the women's clothes bundled atop his donkey; he knew the streets and the residents, could tell a good opportunity from a bad one. As the train rumbled west toward the steeples and minarets of the jagged Constantinople skyline, he questioned the men about their work. Before long, they pulled into Haydarpasha, the capital's stately main terminal.

A grand clock stared out from between its two towers like a Cyclops, casting an eye over the bright Bosphorus Strait, normally dotted with bobbing boats. Even after so many trips, Stepan was still impressed by the sight of the majestic new railway station. Recently built by the Germans, Haydarpasha was part of an ambitious project to increase trade and military might by connecting the rail of the Christian West, specifically in Berlin, with Islamic Baghdad.

Stepan exited the station to the docks, where he caught a ferry to Stamboul, the hub of old Constantinople, just a short boat ride away. To Europeans, the district of Stamboul felt exotic, like something out of

"Aladdin and His Magic Lamp." It was a place of secrets, of city sounds that seemed to follow visitors everywhere, of darkened hamams and their arched entryways. The alleys were imprinted with the city's Muslim present and Christian past, with centuries-old churches, tombs, and mosques peeking above the buildings.

Stepan pushed down the lively streets of Stamboul until he reached Mahmud Pasha Street, the location of his modest boarding house. The narrow road was lined with cramped shops and merchandise carefully arranged under awnings, the names on storefronts written in the various languages of the empire — Armenian, Ottoman Turkish, and Greek. Men with fezzes and women in hijabs (headscarfs) could be seen strolling through the outdoor bazaar amid men in tall hats and women in puffy Western skirts.

That night, Stepan climbed into bed and closed his eyes but couldn't sleep, his wakefulness fueled by excitement. An *emanetji*, he mused. He calculated distances, demographics, baggage costs. The equation was simple, really; if he could withstand the initial losses and convince others to trust him, he would succeed.

Of course, he knew he couldn't endure heavy losses for long. When Stepan was ten years old, his father, Hovhannes, had died unexpectedly, and since then, money had been tight for the Miskjian family. His father had co-owned a successful hardware store, and Stepan had expected to work alongside him one day. With Hovhannes gone, his wife, Hripsime, had had no choice but to take her two boys — Stepan and his older brother, Armenag — out of school immediately and turn them into breadwinners.

Stepan knew the stakes. If he failed as an *emanetji*, his family would lose its livelihood, its *lavash*. His two older sisters, Zaruhi and Aghavni, had married and moved out, but the two younger ones, Arshaluys and Mari, were still at home and very dependent on him. Despite not having much capital, he did possess another form of currency, perhaps more valuable: his reputation in the bazaar. He was rich in this, he told himself, having started his peddler business on his good name alone. At the very least, it would allow him to take out a loan for a job that seemed custom-made for him. Though only five feet four inches tall,

Stepan was strong, and his long legs resembled a spider's, ideal for carrying goods across the web of streets in his *sanjak*. In a month, he would turn twenty-four years old. Did he want to remain a peddler forever? Surely the universe had bigger plans for him than selling goods from behind the haunches of a donkey.

The first morning back in Adabazar after his trip, Stepan rose early in his three-tiered house. He crept past his siblings, Armenag, Arshaluys, and Mari, and didn't divulge a word of his intentions. He kept quiet on the first floor too, since his widowed mother slept not far from the front door, as if to protect them all. Always dressed in black, his mother, Hripsime, was familiar with her younger son's high jinks, and she disciplined him for the most minor infractions, like the time he fashioned one of her fine slipcovers into a canopy for his rickety carriage. What would she think of his new endeavor? Thankfully, he managed to leave without arousing any suspicion, and he stepped out onto Nemcheler Street, which arched its back like a ruffled cat on the edge of town.

Down the avenues, the freshly shaven Stepan walked briskly, his wavy hair parted for business, the wrinkles smoothed out of his clothes. He passed through neighborhoods of two- and three-story dwellings with small flower boxes outside windows and fences. Typical Ottoman architecture, the buildings were made of wood and stone, much like Stepan's own family's home, only his was freestanding. Above, wooded hills hemmed the plain. The town slowly awakened, the workers hurried to the mills and the fields, the horses tugged loaded carts, the Armenian women started to spin yarn on their wheels, and Stepan's own sisters tended to their cocoons of silkworms for extra cash. Almost everywhere else in the empire of some twenty million—a stew of Arabs, Turks, Greeks, Jews, Kurds, and others—the two million Armenians were outnumbered. But in Adabazar, about a hundred miles east of Constantinople, Armenians made up half of the approximately thirty thousand residents. One could easily determine the Miskjians' Armenian ethnicity by their last name: it ended with *–ian*, or "son of."

Like other families, the Miskjians were deeply religious and proud of their heritage. In the early 300s, back when they had a country to

call their own, the Armenians had been the first people to formally accept Christianity as their national religion. The four quarters of their
exalted town were named after the churches that rose up like an altar
in the middle of each one. But the courier-to-be wasn't that observant,
really; he had once schemed to power back shots of *oghi*, an anise-fla-
vored spirit, before service with his friends; and had orchestrated one
of his biggest pranks inside the sacred halls of a church. Still, he'd be
awfully lonely if he skipped the religious events altogether. Spiritual-
ity permeated Ottoman-Armenian life, influencing social gatherings,
schools, and governing councils. To the Armenians, their hometown
wasn't called Adabazar; it was Asdvadzareal Kaghak, the God-Created
City. Never mind the intermittent plagues, the biblical flooding by the
nearby Sakarya River, and the fault line that trembled underground.
The God-Created City was Stepan's home.

After making flyers at his friend Harutiun Atanasian's print shop,
he proceeded to the Uzun Charshi, the Long Market, where Armenag's
store sat tucked away amid the other Armenian-owned shops, bunched
together like a tuft of wool. The Turks, the town's other majority, congregated their business at the bazaar's opposite end, and the Greeks
and the Jews toiled somewhere in between. Nervous, Stepan knew Armenag wouldn't approve of his new venture, but he was duty-bound to
tell his twenty-six-year-old brother about this idea. Given the speed at
which gossip raced from stall to stall, he also couldn't delay.

Armenag appeared in the storefront now, handsome with his thin
mustache and baby face. Though he towered four inches above Stepan, the brothers bore a strong physical resemblance to each other, with
their matching square jaws and deep-set eyes. This was a surprise visit,
yet Stepan didn't say a word of greeting. Instead, he handed Armenag
the newly printed flyer. Curious, his brother read, "'Starting tomorrow I will make a weekly round trip to Constantinople and am ready
to take all types of parcels and goods for pickup or delivery. Those who
want to see me can find me at my friend Baron Mihran Sahagian's coffeehouse.'"

By the last line, Armenag's face had contorted as if he had just
swallowed curdled milk. "Those are *bosh* plans," he sniffed. *Bosh* meant

"empty." "Your present work is good and more secure, and so far we are getting by with honorable work."

Armenag's colleague had overheard the debate and wandered over, wondering about the fuss between brothers. In the close-knit community, butting into others' affairs was a birthright. Armenag passed him the flyer. "Huh," the man said, and then he sided with Armenag. "Those are empty plans. Don't act like an ignorant teenager."

But Stepan knew the timing for this venture was perfect. Just two years earlier, in 1908, a political revolution had drastically changed his life, and that of many others in the empire. Before that Stepan could only sell his goods only within his *sanjak*. After the rebellion, the new leaders lifted the long-standing travel restrictions, and Stepan was now allowed to take trips to Constantinople for business. His profits quickly doubled. The party behind the advancement, the İttihad ve Terakki Jemiyeti, or Committee of Union and Progress (CUP), had blazed into power by seizing government buildings and telegraph lines in the Balkans. Battalions dispatched to quell the rebellion had even switched sides. Fearing a march on Constantinople, the despotic Sultan Abdul Hamid II gave in to their demands and reinstated the constitution, which guaranteed rights to all, regardless of creed or ethnicity. Though the sultan remained in place as a symbol, he had lost all his power. The new Muslim leaders were intellectuals and exiled officials who became known as the Young Turks. Their passion was contagious, and celebrations of their victory had clogged the roads of Adabazar as Armenians proudly waved the Ottoman flag and tossed flowers to welcome the new era of the CUP and its professed trinity of ideals: liberty, equality, fraternity.

To the Armenians, it had seemed like this day would never come. Under Ottoman governance, non-Muslims were broken into *millets,* or religious communities, each with its own patriarch, local councils, and distinct laws that often treated them unfairly. Families like the Miskjians struggled with fewer rights than the Muslim Turks had, paying higher taxes to the government and local officials and lacking a voice in court. They were also denied the right to bear arms, unlike their Muslim neighbors, which left the Armenians vulnerable to attack. They dreamed of a better life. Theirs was a golden history with Armenian kings who reigned over swaths of land in the east and beyond. The eastern provinces of the Ottoman Empire had been their ancestral homeland; their roots dated back thousands of years. However, that was a long time ago, before all the conquests and the lengthy list of rulers: the Persians, the Macedonians, the Byzantines, the Arabs, and, most recently, the Seljuk Turks.

Being an oppressed minority in such a vast empire had taken its toll, especially in the late 1800s. When Stepan was about eight years old, frustration among Armenians mounted. They wanted reforms. The sultan had previously promised to install a new constitution that guaranteed certain rights — but it was quickly suspended. Pockets of Armenians rose up. The sultan cracked down, and many were killed. Nonetheless, still hoping for change, a few more Armenian towns rebelled. The empire's response was swift, and the newspapers covered the tragic consequences. "Another Armenian Holocaust: Five Villages Burned, Five Thousand Persons Made Homeless, and Anti-Christians Organized," read the *New York Times* on September 10, 1895. In all, some two hundred thousand Armenians were massacred from 1884 to 1886, and many were forced to convert to Islam. This violence earned Abdul Hamid II the epithet "the Red Sultan."

Now 1910, the feared sultan was exiled. The year before, in 1909, a group of religious fanatics tried to restore the sultan to his full power, and oust the CUP lawmakers from office. Still, to quell this attempt at an overthrow, the threatened and vulnerable CUP did what would have once been unimaginable: they turned to Armenians for help. No longer barred from military service, the Armenians formed a provisional armed unit to protect the fragile new government. Though the overthrow was thwarted before the militia saw much action, the Armenians were proud of defending their new rights, and the sultan was officially deposed. At last, a progressive government was in charge, and the possibilities for the future seemed endless.

In the bazaar, Stepan wasn't surprised by the shopkeepers' reactions to his new business. *Closed-minded,* he thought as he walked away with his circulars. The town had been founded, after all, on the principle of commerce. It was even incorporated into the town's name: Ada-Bazar, the Island Bazaar. Unswayed by the skeptics, he made his way down the Uzun Charshi, pasting up a flyer at every Armenian coffeehouse. As he papered the commercial district, criticism of his new venture mounted. Traders and customers mobbed him with questions; one critic pronounced, "This type of work won't pass muster in Adabazar!"

Stepan stepped out of the marketplace and brushed off the cynics. For once in his life, he felt absolutely certain about his path, his optimism buoyed by the warming weather. Another harsh winter was behind him; the flowers and plants around town were poking out of the soil, as if no longer afraid to bloom; the deadened trees were bearing their first leaves. He, too, would grow. He, too, would break new ground.

Despite those negative rumblings in the bazaar, the hopeful Stepan was in good company. His greater community was undergoing an unprecedented cultural awakening, ecstatic over their new constitutional rights: freedom of assembly, freedom of the press, and freedom of speech. Cafés were stuffed with bundles of new Armenian-language publications as a mostly skilled working-class populace swooned over all things intellectual and arts-related. The once-repressed citizenry was almost unrecognizable; it had its own reading room, stocked with two hundred books. "The true readers have been checking out books and newspapers with pleasure," read one article in the inaugural issue of *Piûťania.* There were also credit unions funded by Armenian capital, a sports club, and even ads for coloring one's caterpillar-size mustache.

For the rest of the morning, Stepan hawked garments in the Greek quarter, his mind spinning like the water wheel at the edge of town. In all these years of peddling, he'd been reliable, rarely returning home empty-handed, pressing through his fatigue, his faithful donkey by his side. On the days Armenians were in church, he traveled

to the Turkish or Circassian areas; when they were busy, his shouts reverberated through the nearby Greek villages. He gave a warm *yassas,*
"hello," to his Greek customers, a *marhaba* to the Turks, and a *parev* to the
Armenians, striking up new conversations with his rainbow of friends
and clients to land a sale.

However, when the hands on his pocket watch twinned at twelve,
he returned to Mihran's coffee shop and saw the bulletin board with
his sign-up sheet — inked with three names, including Antranig Efendi
Merjanian, one of Adabazar's most respected moneylenders, a man so
revered that everyone addressed him as Efendi, the equivalent of Sir.
His interest was a clear vote of confidence, Stepan told himself, and he
set off to meet his first client.

On seeing Stepan, Antranig Efendi smiled widely and congratu-
lated him on his pioneering endeavor. The key to success was trust, he
said, and Stepan was steeped in it. He pointed to some provisions and
asked the cost to transport them to his sons in Constantinople. Stepan strode across the room, lifted the heavy basket, and told him to
pay whatever he wanted. He did the same with the next two men. At
the station, he was charged twenty-seven ghurush — nearly three times
what his customers had given him. No matter; Adabazar's new courier
was soon transporting his first load to the capital. No one needed to
know it was at a steep loss.

On Stepan's return two days later, his brother, Armenag, was waiting
for him, his soft features sharpened into agitation.

"How did it go?" the older brother asked.

"This time, I only recovered my expenses," he fibbed. "I didn't make
any profit."

"Didn't I tell you that this is an empty plan?" Armenag said.

"I'll succeed soon."

Armenag tried to persuade Stepan to alter his course. Any decision
Stepan made — especially the reckless financial ones — affected all of
them. As the elder man of the family, Armenag felt particularly responsible. They were barely making it, surviving mostly thanks to the thrift
of their mother. Armenag knew he'd be remiss if he didn't say anything.

"You know we won't be able to live on the wages I make," he stressed. "What will happen to us if you don't make any profits?"

"I'll continue my street peddling two days a week, so I'll be able to contribute to our family expenses while I build this other business," Stepan told his brother.

In reality, he had already stopped his peddling. He was living exclusively on loans now, and in just one week, he had lost fifty-five ghurush, nearly half a Turkish pound, eleven times what he used to earn in one day.

During a trip to Constantinople for business that year, Stepan set out by tram to the Hill of Liberty, high above the city. He and a friend were attending a ceremony commemorating the Muslim and Christian martyrs of the revolution, buried together in one grave. These were the men who had halted the countercoup of 1909, who had given their lives for their new liberties. Thousands of people, including dignitaries draped in elegant dress, crowded around Stepan to honor the heroes. Afterward, with the tram no longer running for the day, he and his companion walked back down the hill, surrounded by the city's fashion-minded Christian and Muslim elite who were conversing in German, French, Ottoman Turkish, and Italian. Their worldliness suddenly gave Stepan an idea. He signaled to his friend, and the two began speaking a made-up language that they routinely practiced, combining foreign words and talking loudly as they strolled. *"Ari chepisheh, medareh sefish, che meh khosheh, part ghoch,"* they said, the sentence part Armenian, part Ottoman Turkish, part Stepan. Immediately, the learned people turned toward these two overlooked men, taking guesses as to their origin. Stepan wanted to laugh but kept his face perfectly straight.

With all these visits to the capital over many months, Stepan was starting to project an image of success back home, but that did nothing for his bottom line. His losses proceeded to spiral. No longer able to haul shipments by hand, he was paying out more and more to porters, cart owners, and the railroad.

For an entire year, he had to float the business like this. By 1911, his debt had climbed to one hundred gold pounds. "Enough losses!"

he finally said to himself. "I'll set my own prices." The results of that revealed themselves in just one week: Stepan broke even. The next week, he actually earned money. Soon after, he started bringing in eight to twelve pounds a month, and he expanded into cash transfers by undercutting the official post office.

No longer a peddler, he was a bona fide *emanetji*.

He was a part of the town's renaissance. All around him, people faced off in political debates and even discussed the merits of women's rights. This was unorthodox and groundbreaking for a patriarchal culture in which a new bride couldn't utter a peep to her father-in-law until he gave her the green light. But change was in the air: a thousand Armenians were congregating together, largely because they finally could. Schools were too small for them, as were the churches and even the old silkworm mills. Stepan wanted to remedy this. The town needed a place for its Armenian citizens to assemble. The auditorium should be modern, Stepan thought, and large enough to fit two thousand people, with a stage, chairs, the works. To realize this dream, he joined a community group, and together, the dozens of men began to raise money for the hall.

Even when their country went to war against Italy, they continued to plan for the building's construction. At a play that December, as the evening performance began, the political tensions of the time surfaced. Two drunken men punched the ticket collector and entered the chamber where Harutiun Atanasian, the erudite publisher of the new Armenian press, was seated. "Quiet down," admonished Harutiun, prim as ever in a dapper suit and round glasses. Immediately, one of the men grabbed the intellectual's hat and threw it to the ground, a burst of drama more mesmerizing than that unfolding on the stage. "He has put on an Italian hat. What purpose is there for an Italian here?" the man slurred, and then he uttered his opinion of the Italians, "Those bastards!"

The drunk expressed what many were feeling. Just a few months earlier, the Italians had attacked western Tripoli, the Ottomans' last real North African territory, given that Egypt had practically become British. With that, the Turco-Italian war commenced, and it dragged on into 1912. Then the Dardanelles came under fire too, so the Turks closed the slender passage joining the Aegean Sea and the Sea of Marmara, resulting in a massive aquatic traffic jam. Next, a handful of Ottoman islands in the Aegean Sea turned Italian almost overnight. In the press, the empire was called "the Sick Man of Europe," as it kept losing land. Several years earlier, Bulgaria had announced its autonomy, and Austria-Hungary had annexed Bosnia-Herzegovina. In fact, starting with the independence of Greece in 1830, the Ottomans had been hemorrhaging territorial possessions in Europe, which they had viewed as the core of their empire for hundreds of years. Their boundaries still stretched from the Balkans all the way across Mesopotamia to Arabia, but the tension in the region was rising, and even the fortunes of the Young Turks had flipped. After a few years in power, they lost their political support and positions in the cabinet.

Stepan paid attention to the news, but he was more focused on the road in front of him. Up and down the seven hills of Constantinople he traipsed, across streets that seemed to angle up at ninety degrees, his bundled packages over his shoulder. The other couriers competed with him, but he edged them out easily, especially as some of them stole the goods in their care.

A posh banker in the capital was shocked when he entered the boarding house in Stamboul and finally saw the wizard behind the money flow: Stepan, the now twenty-six-year-old, five-foot-four-inch former peddler of Adabazar. "I congratulate you, my son," he said. "You'll become a great man."

Stepan believed this. He never imagined what was to come — that world events would soon ripple to his doorstep, that as the war with the Italians ended, another one would begin in the Balkans that October, and that his beloved older brother, Armenag, would be drafted.

Armenag said goodbye, left home, and set off to register. He wrote once from his post in Adrianople and then no more; soon the Bulgarians surrounded the Ottoman provincial capital, trapping him inside.

The Countdown

2007

"I'M GOING TO TURKEY."

My mother's face fell. "Why?"

She was sitting across from me at the dinner table. Always the health-conscious cook, she had made steamed broccoli, asparagus, soy burgers, and bulgur and had heaped hefty portions onto my plate. She was now waiting for me to eat.

Instead, I continued. "I want to follow Baba's footsteps. I want to see where he went after leaving Adabazar." I let the moment rest, forked a green spear, then trod forward carefully. "And I also want to go to Deir Zor."

My mother stared at me, her eyes betraying fear. That last word echoed, much like *Auschwitz* does to the families of Holocaust victims and survivors. It was the terminus of the death march, of all the encampments that had once dotted present-day Turkey and Syria. This was the place her father called hell when he spoke of it, and for good reason. The region around Deir Zor was one of the empire's worst slaughterhouses. These real-life horror tales were my mother's bedtime stories. When referring to it, as in "No one made it out of Deir Zor" and "People only died in Deir Zor," she'd drop her voice as if lightning were about to strike. My mother was visibly upset by my announcement. This was unusual — long ago, she'd picked up that child-of-war ability to hide her worry. "Don't go," she said.

But nothing could convince me otherwise. "Mom, I am going," I declared. She averted her gaze, knowing that I wasn't a kid anymore and that she couldn't stop me.

Soon after, I booked my ticket for the upcoming summer. I wanted to travel there during the intense heat to better understand the weather conditions my grandfather faced. I had only a few months before my departure, and I jumped into planning mode. I had to plot my route, find a Turkish fixer who would not be offended by my subject matter, an Arabic translator, and a driver to guide me across Syria where my grandfather had walked. Aching from four immunizations, I steadily made my way down a long list: flashlight, camera, notebooks, backpack, thumb drives, long-distance phone card, laundry detergent, skin ointments, antibiotics, anti-inflammatories, and conservative clothing — the last item not easy to find in the beach culture of Los Angeles, as I learned one afternoon in Pasadena. Looking for a long, conservative skirt, I traipsed up and down the main thoroughfare and ducked into all the usual storefronts — Banana Republic, French Connection, J. Crew — with no luck. I asked at half a dozen shops, each time receiving a polite "No."

Finally, at some small, unmemorable boutique where auto-tuned voices whined through the speakers, I received a different response. "Oh, yes," answered the young woman, no more than twenty, as she went to the back. Hurriedly, I followed her, watching her high blond ponytail sway with her steps. She dug into a circular rack and pulled out a garment for me to view. I almost laughed; the blue fabric stopped two inches above the knee. "Where is the long skirt?" I asked.

"This *is* long," she said.

In preparation for my August trip, I read as many accounts of World War I and the Armenian genocide as possible. As I reviewed countless survivor oral histories, I began to see just how painstaking my grandfather's journals were. Those chronicles were eerily similar to his, only with different names and varying levels of detail.

The people from Adabazar who told their stories often returned to the same defining events before the war as my grandfather did, like the ill-fated parade to celebrate the Armenian alphabet. Wanting to know more about that — and everything else — I inevitably wound up at Abril Books in nearby Glendale (dubbed "Armendale," for its large Armenian population). The storeowner, Harout, always helpful and with a welcoming husky voice, pointed me to the bookcases filled with genocide-related material in both English and Armenian: newspaper clippings, consular reports, missionary telegrams. The number of dispatches chronicling the killings during that time had surprised me; given all these primary sources, I didn't understand how there could still be debate, how the Turkish government could deny that it had happened, how educated Turks could espouse their government's position and, appallingly, blame the victims themselves for their deaths. I spent hours crouched behind the bookshelves at Abril Books. On a May day three months before my departure, I showed up and discovered Harout sitting in the back office drinking thimble-size portions of Armenian coffee with two men, one of whom was wearing a black shirt with a white clerical collar.

Harout set down his cup and waved for me to join them. Then he said loudly, as if introducing a stage act, "This woman is writing a book on the genocide. And she is following her grandfather's path, traveling from Adabazar to Deir Zor!"

Surprised, the men looked at me and smiled. "Are you planning on doing this alone?" the priest asked.

"Yes, but I was looking for someone I could hire to help me."

"I can help you. I have traveled there before and know people. Don't talk about it with just anyone. That would be dangerous."

I was grateful for the priest's offer, and we exchanged contact information.

Not long afterward, during a lunch with an Armenian academic, I was again reminded of the risk. When our conversation turned to my upcoming travels, the man's face became very solemn; his tone stiffened. "Don't go," he said. "We don't need another martyr."

His warning scared me, but I knew that I couldn't cancel the trip, even though his concern was valid. Only months earlier, Hrant Dink, an Armenian-Turkish journalist, had been murdered in Istanbul. I remember how stunned I'd felt when reading the *New York Times'* coverage, and I hadn't even known him, only his work. "A prominent newspaper editor, columnist and voice for Turkey's ethnic Armenians who was prosecuted for challenging the official Turkish version of the 1915 Armenian genocide, was shot dead as he left his office," the story read.

I'd stared at the images online: a rumpled white sheet covering his body, the bloodstains visible. Why was history repeating itself? It had been nearly a century, yet the enmity seemed the same. The assassin—seventeen years old, just a child—was later lauded as a hero by nationalists. Police even posed for pictures with him after his arrest, the boy holding the Turkish flag. The teen's supposed motive: to stop Hrant from ever speaking about the genocide again.

Just before the murder, Hrant Dink had been convicted of "insulting Turkishness." Article 301, a relatively new Turkish penal code designed to silence a nascent dialogue about the past, had made this a crime. Other Turkish writers, historians, and intellectuals had similarly spoken out and been charged, including one of Turkey's most famous sons, Orhan Pamuk, the recipient of the 2006 Nobel Prize in Literature. Pamuk's alleged offense was telling a newspaper reporter this: "A million Armenians and 30,000 Kurds were killed in this country and I'm the only one who dares to talk about it."

I needed to navigate this landscape carefully. To maintain a low profile on my journey, I planned to stress my American nationality in Turkey and emphasize my Armenian one in Syria, where a large and respected population thrived. I'd use the words of the Turks in Turkey, referring to that moment in history as the "Armenian question." But I would not use their other moniker, "the so-called Armenian Genocide."

"You're not going to tell people what you are doing, Dawn, right?" my mother asked repeatedly.

"Of course not," I always said.

I lied.

In Abril Books that May day, I bade goodbye to the priest and headed toward the door. At an island of books, I paused to thumb through one that was filled with maps of the Ottoman Empire. A handsome man with short hair walked in. He stopped nearby. "Are you thinking about buying that book?" he asked.

"Yes," I said, immediately looking down, suddenly shy. *Why did I do that?* I chastised myself. At the register, I set the book on the counter, and the man stepped into line behind me.

Ringing up my purchases, including the map book, Harout looked back and forth between the striking young Armenian man and me. His smile widened. "Are you a bachelor?"

"Yes," the man said.

Then Harout turned to me. "And you?"

My face reddened in embarrassment at the Armenian matchmaking. "No," I replied, keeping a straight face, like my grandfather. "I'm not a bachelor."

Once home, I opened up my map book and hastily began to plot my route across Turkey to the Syrian border. I traced my finger across the pages, along the train tracks, mountain ranges, and stretches of desert, and memorized my way, soon to chart this path for real.

Alphabet City

1913-1914

FOR THE MISKJIANS, agonizing months had passed without any news from Armenag. The forlorn Hripsime sat outside spinning wool, her long, thin face longer still. Stepan reassured his mother and sisters that Armenag would survive the war and come home, despite not believing it himself. Then, in the summer of 1913, Armenag returned.

When he arrived by carriage, Hripsime raced out of the house and embraced her starved and shaken son, holding him close, which she hadn't been able to do in nearly a year.

Armenag told his family of his trying ordeal in the Balkan war — the cold winter he'd spent fighting for the Ottoman Empire, his subsequent capture by the Bulgarians as a prisoner of war. He'd been so hungry during the siege of Adrianople, he and the other soldiers had dug up the decaying corpse of a horse for food.

The Miskjian family was intact, but the empire was not. The sidelined CUP — which had lost its political will — had criticized the proposed armistice to end the Balkan war and used it to stage a coup d'état. This led to a second short war that, mercifully, concluded soon afterward.

During these wars waged against Christian countries, the Ottomans suffered tremendous losses. They ceded most of their European territories; and countless displaced Muslim families poured into Anatolia in search of new homes. Increasingly, the Muslim majority, who had previously embraced the ethnic and religious inclusiveness espoused by the 1908 revolution, began eyeing their Christian neighbors with hatred. All this further destabilized the already shaky region and set the stage for more turmoil.

For the Miskjians, though, the only thing that mattered was that Armenag had come home. Eager to resume his life, Armenag teamed up with Stepan, who was now several years into his courier business and earning six pounds a week. Stepan soon passed out letterhead with a brand-new name: Miskjian Bros. Depositary Service.

The brothers traded off doing deliveries and paperwork, and Stepan was thrilled to have Armenag by his side. The schedule allowed them both to rest and devote time to other activities. Stepan and the other members of the community group had raised enough money to purchase land for the assembly hall. Board by board, they were building a house for their future as the townspeople around them prepared for an even bigger event: the fifteen-hundred-year anniversary of the creation of the Armenian alphabet. All over the *vilayets,* provinces, the Armenians were about to mark the moment fifteen centuries earlier when two holy visionaries, Sahag and Mesrop Mashdotz, created the Armenian script. Similar to Latin, the Indo-European text was read from left to right, unlike the Ottoman Turkish Perso-Arabic characters, which flowed in the opposite direction. Along with religion, the script was the foundation of Armenian identity, allowing for the documentation of hardships and triumphs, history and love. "The Celebration of Armenian Civilization," blithely declared *Stamboul,* a French-language newspaper in Constantinople.

The Adabazar festivities would begin on October 26. For the jubilee, the newspaper was predicting parade-perfect weather: a crisp fifty-seven degrees and a "beautifully calm" day at a time when the psychological temperature of the empire itself was boiling hot.

When that clear autumn morning finally arrived, the Adabazartsis made their way toward the slender bell tower and entered the Surp Hreshdagabed Church with its many stained-glass windows. Usually,

Sunday services drew only a modest attendance, but for this event, finding space required the kind of divine illumination that had created the alphabet in the first place.

In front of the energized masses, an elderly archbishop appeared and began his homily. Stepan recognized the lines of his craggy face — at least, what he could see of it. A triangular hood sloped down from the top of his head to his bushy eyebrows; just beneath his nose, an unruly snow-white beard curled onto his chest. It was Stepannos Hovagimian, the foremost Apostolic authority of Izmid *sanjak*.

Following the inspiring homilies, the attendees spilled into the courtyard, where a stage had been constructed underneath the branches of a gargantuan pine tree. For the next two hours, men took to the podium, their voices overlaid by the beat of the band.

On that bright fall morning, young children were the first to parade through the streets. Dressed in lavender, eighteen pairs of boys and girls proudly led the way, passing the joyous crowds and neat arrangements of bright flowers, their tiny hands holding flags of different letters high above their heads, an Ա, Բ, Գ, like a moving game of Scrabble.

Students followed, twenty to forty from each of the town's eight Armenian schools, the thousands marching together along the parade route, clutching banners. They funneled into an alleyway and stopped in front of the headquarters of the Tashnagtsutiun (Tashnag), the political party Stepan had recently joined. There, the crowd listened to a rousing speech from one of the party's members: "We believe that this holiday represents all Armenians," exclaimed Dikran Baghdigian, a teacher, "and thus rightfully we call this holiday a national one."

His talk briefly touched on the controversial subject of Russia, a bitter enemy of the Ottoman state. Most of the Ottoman Armenians resided in the six *vilayets* in northeastern Anatolia, the others in nearby Russia. This was the ancestral homeland of Armenians, divided by the Ottoman-Russian border. In fact, it was Russian-Armenians who had established Dikran's party, the Tashnag, or Armenian Revolutionary Federation, which operated in both countries with the common goal of improving the lives of all Armenians, wherever they were. Over in Russia, the Armenians had contributed to a general fund, Dikran explained, despite restrictions from their government, adding, "Let's follow their example and openly donate to the fund so that we may also reach our goal."

This kind of Armenian nationalist speech roused the crowd. To some Turks listening, the statement seemed radical, even treacherous; the suspicions heightened since the Tashnag Party had stopped supporting the Young Turk government just before the Balkan Wars. This was a particularly delicate time for the Muslim Turks, as they had already watched so many *giaours,* or infidels, secede over the years. The empire was especially on edge because of the Balkan defeat, and Turkish nationalism was flourishing; the government would preserve the rest of its shrinking territory at all costs. Some Turks even went so far as to suggest that in self-protection, the empire had to rid itself of its non-Muslims or, as one official called them, its "internal tumors." The discord grew when the Armenians, still unhappy with their second-tier status, pressed for reforms again.

From the podium, Dikran seemed to be calling for a unified Armenian community, exactly what the Turks feared most. The Turkish attendees of the parade looked anew at the alphabet festooned on the flags. It was a coat of arms, they decided. The whole day was a procession of politics and misplaced nationalism, not a celebration of culture. In that alphabet, the Turks saw an ominous message. Concern reverberated throughout the empire's ornate halls of power after other leaders attended similar celebrations in the capital. "The national demonstrations of the Armenians in Constantinople began to take unusual forms," wrote Jemal Pasha, then the city's military governor, in his memoir. "They organized great celebrations in memory of the anniversary . . . of the discovery of the Armenian alphabet. They carried their audacity to the point of throwing confetti in the Armenian national colors about the streets. We bore all this with unshakable patience, and took the necessary steps to prevent ugly incidents."

From the Tashnag building in Adabazar, the parade merrily pushed forward. At precisely three o'clock, a Turkish police lieutenant and a handful of soldiers burst onto the scene. "The march is forbidden," the deputy thundered, saying that the organizers hadn't obtained permission. "People should disperse."

The paraders swayed to a standstill, and the band members lowered their instruments. The jamboree was far from over; there were many more stations to go, including the site of that evening's grand finale, the Manishag School, where, to the sweet sounds of singing, thirty-six lanterns would light up the kindergarten, each lantern representing a letter of the original script.

Now Krikor Sukerian, an Armenian lawyer in attendance, wedged his lean body into the unmoving crowd to mediate. "The parade is a peaceful procession by school pupils in Armenian quarters and doesn't have a provocative or peace-disturbing nature. There's no reason to prevent it," he said.

The lieutenant remained unswayed. Armenian elders huddled briefly and then rendered their decision. "Continue playing," they ordered the band defiantly. "And move forward."

In response, the lieutenant told his men to ready their bayonets. Knocking down flower arrangements, the Turkish soldiers loaded their weapons and took aim at the students as petals fell to the ground. The horns blared and the drums tapped, the atmosphere as tense as a tendon about to snap.

"Fire!"

On the sidelines, Stepan watched the frenzy unfold: the people fleeing, piling on top of one another. The trampling feet, the screams of the adults, and the higher-pitched shrieks of the children caught in the middle of it all would have resounded in his ears. A group of Armenian men, including a uniformed police officer, sprang toward the soldiers, fearing that if they waited another moment, it would cost little lives. They wrestled with the soldiers, trying to pry their weapons away.

From this struggle, the lieutenant emerged, his face crimson with blood. At three o'clock, he aimed his gun upward and pulled the trigger, summoning his unit. All his men except one materialized. The officer's eyes searched until he located the missing soldier, his body pinned down by a young Armenian man named Ashod Tumayian. The lieutenant couldn't believe what he saw. Stepan Miskjian would always return to this image as a defining one: The soldier was dead. Stabbed, supine, still.

Up ahead, the children proceeded to hold their alphabet flags in the air. As they were hurried away, one boy wailed, "I don't have my *P*" — he'd dropped it somewhere along the way.

By morning, the Turkish press was aflame. "Met with blows of batons, canes, and daggers, and slaps, those soldiers did not even take the initiative to defend their own lives with their own weapons," criticized *Tanin*. "If they had wanted, they would have been able to injure any one person. Even when calling for help, the militant leader did not point his gun towards any assailant. It was the sacrifice of those soldiers' lives that will leave a shameful mark upon Adabazar."

In the forty-eight hours that followed, Armenian teachers, priests, community leaders, and musicians were arrested, including the political speaker Dikran and the lawyer Krikor. Armenian businesses were closed. A curfew was imposed. News reports told of troops arriving to calm a nervous populace; arms were confiscated and one hundred and fifty Armenian homes searched. Only licensed hunting rifles and posters of the Armenian political parties were discovered, but they were seized anyway. Soon after, criminal proceedings in the soldier's death commenced. While Dikran, Krikor, and many others were released, the court found Ashod Tumayian guilty of murder, and it sentenced Aram, the Armenian police officer, to a decade behind bars. A host of others involved in the milieu, such as the local butcher Bedros Giuzelian, received lighter sentences.

On a sorrowful day in late October, the fallen Turkish soldier was laid to rest. Thousands attended the burial, including his military colleagues and former classmates. A red flag with a white crescent and star draped his casket. And in his memory, a monument was erected, inscribed as follows:

> During the course of my duties as a soldier
> Armenian *giaours* cruelly slashed my throat,
> Oh Moslem brethren, whenever and wherever you can,
> Don't forget to avenge me.

· · ·

By December, the town seemed to breathe again. Around that time, a renowned gun dealer, Baron Sisag Chilingirian, entered the Miskjian brothers' office. The baron explained that he had bought some weapons from a wholesaler in Constantinople and needed them in time for the big Monday bazaar.

Hastily, Stepan made the trip to the capital and submitted the baron's order. When he picked up the goods the next day, Stepan asked, just to be safe: "Is there anything prohibited in the shipment?"

"No," the man replied, "everything in the shipment is commercially available and sold on the open market."

Stepan hauled the rifles, cartridges, and other packages to the station and purchased a ticket. On the way home, the train made quick stops in Pendik, Tuzla, Gebze, Diliskelesi, and Tavshanjil. At Hereki, the train idled a little longer, allowing Stepan to briefly get off and stretch his legs. When he heard the bells chime, he made his way back to the compartment and discovered a policeman blocking the entrance.

"Are the parcels in this car yours?"

"Yes, they are."

"What are they?"

"Commercial goods," Stepan explained and he drew out the invoice to show him.

"I don't need to see it." The officer addressed the passengers in the seats beside Stepan's: "Keep watch on him."

Believing the encased items were military guns, the Turkish travelers in the compartment had informed the authorities there was a suspicious Armenian in their midst. Though Stepan knew he had done nothing wrong, he felt increasingly uneasy. A monumental misunderstanding had occurred. He feared arrest.

Twenty minutes later, after the train passed the serene lake of Sabanja, the normally reassuring sign of home appeared. Stepan's eyes fell on the Adabazar station. A mass of officers, half of them policemen, half gendarmes, were crouched in shooting positions, their guns leveled at the train. In fact, soldiers guarded the premises everywhere. In just a few minutes, the sleepy two-story terminal had been transformed into a military staging zone.

A voice shouted for all passengers to remain in their seats. On the platform, the soldiers fanned out, hunting for a specific car, *his* car, the number of which had been cabled over from Hereki.

"He's here!" yelled the riders beside Stepan.

The police signaled for everyone else to disembark. As the travelers hustled off, the officers collected all Stepan's packages, including the items he'd picked up from other vendors, and stacked them in a corner. Ten minutes later, the station was empty except for Stepan and the armed guards. To think, several months earlier, Stepan's biggest problem had been a snafu in the delivery of children's socks. The chief now tore open one of Stepan's boxes and peeked inside at the piles of hunting rifles. The man seemed disappointed; nothing there was prohibited. Where were the illegal Martini-Henry weapons he'd been told about?

Stepan dug the invoice out of his pocket and handed it to the chief, who looked it over and then combed through the remaining parcels. His annoyance increased by the moment, and when he reached the bottom of the boxes, his mouth and nose twisted like a wrung towel. There was nothing more to check. Nonetheless, as he turned to leave, he instructed his men, "Confiscate the packages and take him to the station."

It did not take them long to reach the police station, which was located across from city hall. On the top floor of the compact building, the police *commissaire* pointed to the rifles and grilled Stepan anew. "What are these?"

"The gun dealer Sisag Chilingirian gave me an order to pick up a shipment from a wholesaler. I went and brought it."

Again the police ripped through his containers, not revealing anything new.

"Where does Sisag Chilingirian live?"

"Near Karaaghach Dibi." Sisag resided not far away, near the Armenian quarter. The name — which meant "under the black tree" — stemmed from a large arbor that once stood in the neighborhood. Within half an hour, authorities had dragged in and interrogated Baron Sisag. Since the baron's answers matched those of the courier, the *commissaire* released Stepan. By the time he arrived home, however, Stepan's distressed family had already heard the news. He sought to reassure them, telling them it was a mix-up. But the next day, he learned that a policeman had attempted to extort Baron Sisag: the goods could be released — for a pound. On principle, the gun dealer had refused, and a week later, Stepan found himself back at the police station.

"As things stand now, I've sent the packages that you were transporting to the military tribunal in Izmid," the *commissaire* said. "You've been called upon to appear in front of them, and you have to go tomorrow. There's nothing to worry about since you didn't carry anything banned. They'll question you and that'll be the end of it."

Stepan appeared in front of the military tribunal the next day, and he left the proceedings feeling more confident. *Government red tape,* he thought on his way home. But several months later, in early 1914, a subpoena arrived at his house, demanding his testimony at the Constantinople military tribunal. The date was set for Friday, just two days away. When Stepan caught the train out, his nerves jangled like a coin belt. In the capital, he headed straight to the weapons wholesaler, hoping to glean more information on the case. An employee excitedly revealed that two insiders had said Stepan would be released after the hearing, as would the guns.

At twelve thirty on that momentous afternoon, Stepan reached the military courthouse. Half an hour later, a policeman escorted the courier into a courtroom with eight people and the judge, dressed in a plain black robe. There were his boxes, flaps open on the top. A volley of questions were shot at him over the next hour, and Stepan answered each one the same as he had before. Finally, he was led out so the panel could deliberate. The exchange was so heated that he could hear the shouting from his seat outside. "Under what type of law could you penalize him, since the goods are not prohibited and you are going to return them anyway," one man demanded, followed by another voice: "What is the offense of this man?"

The bickering persisted for another hour. When the courtroom fell silent, the guard brought Stepan back in. A decision had been reached. The magistrate read the verdict as Stepan strained to understand the legalese. Petrified, he made out the words "and under this law, you are hereby sentenced to incarceration and a fine of five pounds."

"Is there such a law in which you could sentence me to jail and fines for a simple transaction of commercial goods?" Stepan pled.

"Silence! Silence!" several voices replied at once. "This is not an ordinary court."

Two gendarmes materialized with chains. After snaking shackles around Stepan's ankles and wrists, they led him outside, their bayonet-mounted rifles held high. Starting in the district of Beyazıt, he was shuffled toward the Golden Horn, past the Grand Bazaar and its stacks of textiles, carpets, and jewels. The patrons stopped what they were doing to stare at the small tethered man flanked by an entourage of officers. *What a great honor to have a guard like this,* Stepan thought, trying to retain his sense of humor. Even from a distance, he could feel the onlookers' terror. *They think I've killed at least a few people.* As always, because of his visibility as the local courier, he encountered some individuals he knew. They lifted their arms upward in that questioning way: *What is happening?* "Nothing," Stepan mouthed, and he released an anxious laugh.

The three-hundred-year-old Blue Mosque, one of the city's most famous landmarks, came into view, its six minarets pencil straight. His destination drew near. He could see it on the other side of the Hippodrome: the imposing fortress of the Sultanahmet Central Prison. The unfairness of his sentencing smacked him like a stone. He was going to jail, and his hard-earned business would suffer, his family too, all because of the alphabet parade. *I've become a scapegoat,* he thought, and he took another step forward.

In an office near the Central Prison two weeks before Easter of 1914, a guard directed Stepan to extend his finger and dip it into ink. Stepan pressed down, leaving his fingerprint. Next, he readied himself for the mug shots, and the camera snapped him from various angles, capturing his aquiline nose, square jaw, large ears. The gates to the building opened. Crossing the threshold, he felt he was entering some alien world. The curious inmates stepped from the shadows and approached: "What is the length of your sentence?"

"Two months."

The answer seemed to disappoint them.

"Two months? Eh, you have come to drink a cup of coffee."

Stepan ventured deeper into the complex, anchored by two large yards, and saw the decrepit barracks, dating back hundreds of years, the rooms stuffed with prisoners, their legs folded as they sat on the floor. Searching for a familiar face, he soon found men from his town, and they took him under their wing. In their communal cell, they made space for him, the convicts often sleeping in lines.

It didn't take Stepan long to make the acquaintance of İbrahim Bey, the much-feared warden. He'd patrol the halls and peer into the men's chambers. Sometimes, he would randomly appear and begin striking Stepan with a large wooden board; the other prisoners would hear the thud of wet planks against flesh. İbrahim Bey deployed a notorious kind of Ottoman torture called *falaka*, or bastinado, beating the bare soles of the inmates' feet until the men could no longer walk. Slowly, Stepan grew accustomed to the sporadic attacks, the nights on the ground, the loss of his freedom. His life on the outside began to float through his thoughts as if it had been a dream. His first customer as an *emanetji*. His brother joining his business. His first cart. The moments felt unreal, imaginary.

Over those weeks, Stepan grew close to some of his fellow prisoners. There were a thousand or so, mostly Turks, with a smaller community of Armenians and Greeks. All clung to one another as if they were dear old friends. They passed the time by telling stories of their lives, taking turns in seated circles. About twenty of them were slated to die by hanging. Fifty men faced life sentences, a condemnation known as "101 years." And the rest had terms of five to twenty years. Now Stepan understood why his stay seemed like a coffee break to the others. He spent many of his hours in the reading room, fascinated by the books and newspapers from around the globe. He was in awe of this collection and taught himself the unfamiliar concepts and words. He especially loved the law books. It was a luxury for an uneducated man like him, pulled out of school when his father died. From this adversity, he had learned to keep his spirits high. *Prison is like a school,* he thought as he buried himself in yet another text.

Two weeks into his incarceration, Easter arrived. Stepan looked forward to Sundays at the prison because an Armenian priest visited the small chapel on the grounds, but no amount of prayers could lift his sorrow on this holiday. Normally, he and his family would be gathered around the table eating his mother's *meze* and *sujuk,* her dried spicy sausage, or else he'd be caroling in the streets with a singing group, raising more money for the assembly hall. Now he sat there, away from his family and everyone else for the next two months, the first of many holidays he'd spend alone.

PART TWO

The Exile

Breaking Stones

1914–1915

STEPAN COULD HEAR A FAINT TAPPING slowly rising above his friends' voices, the sound filling the silkworm mill. Without pausing to investigate, the committee members continued to divvy up the meeting's tasks. Soon the mill would reopen as Adabazar's new assembly hall. Theater! Music! Lectures! Months after leaving prison, Stepan could hardly wait. But at ten o'clock on that August morning, the noise intensified.

The dozens of Armenian men scrambled to the windows and peered out. Through the glass, they could see a town crier entering the main thoroughfare, beating a large double-headed drum, the sticks swinging from side to side. Halting suddenly, he began to shout with all his might, "The government has legislated a general mobilization. As of tomorrow, all men — Turks and non-Turks alike — between the ages of twenty and forty-five will be called up." The messenger started thumping the drum again, the incessant beat further pounding in news about the war in Europe. *Paht-tah-tat-tat. Paht-tah-tat-tat.* Stepan and his friends rushed around, frantic, crying out, "What does this all mean?"

"It's a government order," the town crier said with a smirk. *He seems excited,* Stepan thought as he watched the man spin on his heels and turn toward the Armenian district, the noise fading as he walked. Stepan and the others reeled with confusion, almost dizzy from all the events of the past month. Did this mean the Ottomans were entering the war? The crier hadn't said. Solemnly, the throngs outside began to disperse, treading through muted streets toward home.

The chaos had been set in motion on June 28, 1914, a mere five weeks earlier. Riding in a carriage through Sarajevo, the heir to the Austria-Hungary throne, Archduke Franz Ferdinand, and his wife, Sophie, had been driven into the line of fire of a Yugoslav nationalist. It was a shot that echoed across borders. Years of tension and enmity, exacerbated by the recent Balkan Wars, had primed the continent to blow. Following the assassination of the archduke, Austria-Hungary secured Germany's support and waged war on Serbia. Russia mobilized in Serbia's defense, and, just like that, Europe marched into war.

So far, the Ottomans had remained on the sidelines. The current mobilization of men was said to be precautionary, in case of an attack. However, unbeknownst to Stepan and most of the empire's citizens, the day before the draft decree, the CUP leaders had signed a secret treaty with the Germans.

When Stepan reached his house now, he huddled with his mother and siblings. He could still hear the crier's words outlining the parameters of the draft. At twenty-eight years old, he knew it included both him and his thirty-year-old brother, Armenag. Barely a year had passed since Armenag had returned from the disastrous Balkan war. The family agreed that Armenag could not go this time, could not endure such a trauma again. He would have to pay the *bedel*, the forty-two-gold-pound fee, exempting him from mandatory service. Usually only the rich or those with savings, like Armenag, could afford it. Nevertheless, the less affluent tried; Stepan could see them everywhere, carting around their cherished possessions, frantic for a sale. Armenag urged his younger brother to do the same as him. "Secure a *bedel*," he advised.

But Stepan resisted. Even if he attempted to obtain an exemption, he knew it would never be granted because of his conviction. He prepared himself for what lay ahead. The tenor of the town had already changed. Stepan no longer heard singing or laughter anywhere, the sound disappearing up chimneys like smoke. Around the dinner tables of Adabazar, families clustered together, worried about their departing sons and fathers, wondering whether they would ever see them again. Through the somber night, Stepan lay awake, listening to the silence of his beloved town, his mind buzzing, not knowing if these were his last hours at home. In the quiet bustle of the new day, Stepan set off to enlist. The avenues overflowed with men, rolling in from different quarters like waves. He joined them, all surging toward the same governmental building in the town's center, where a crowd already swelled. Because all the age groups had been called at once, long lines looped through the entire empire, and men stood in the summer heat, sometimes for days on end. The queues for bread also felt interminable, as the recruits hustled to procure provisions.

News of the *seferberlik*, or mobilization, splashed across newspapers, inking in black what most already knew. "The Ottoman government instated a draft," reported the Constantinople newspaper *Puzantion* that August. "All soldiers are required to report to their respective branches within three days. Those who do not respond to this notice will be punished." Evasion meant the death penalty. Still, some men tried anyway, drinking hookah water, hoping to look jaundiced, as the global strife escalated by the day.

Just as the United States declared its neutrality, the Germans invaded neutral Belgium in order to reach France. "Paris for lunch, dinner in St. Petersburg," the German kaiser had announced. But the move backfired, and Belgium's powerful ally, Great Britain, was drawn into the conflict. A cascade of allegiances and entanglements followed. Soon, the Ottomans would sign another secret treaty, this time with Bulgaria. The coalition of the Ottoman Empire, Austria-Hungary, and Germany would grow into the Central Powers. They would fight the Triple Entente, or the Allied Powers — Great Britain, France, and Russia — in what would become the deadliest conflict of modern times.

After Stepan enlisted in Adabazar, he returned home briefly, embraced his mother, brother, and sisters, and said his goodbyes. Swinging his bag, heavy with supplies, onto his shoulder, he left his house and his family. He passed his neighbors and favorite shops, every corner colored by memories. "I will return soon," he told himself. At the designated meeting place, hundreds of draftees had already assembled. Wading into the hordes, he recognized many, since this infantry was mostly Armenian.

A retired captain appeared in front of the recruits. *He's ancient,* Stepan thought, lifting his arm to salute. Standing there, Stepan did not look the part of an Ottoman serviceman with the traditional khaki uniform and pointy helmet. He was dressed in his own garments, like the rest of his lowly *amele taburu,* his Armenian labor battalion. After lining up, Stepan and his company broke into groups of twelve, each overseen by a corporal.

Soon, a cart rolled up. It was filled with pickaxes and shovels, the tools of roadwork, not the guns and rifles of battle. Along with the Greeks, the Armenians were consigned to be pack animals — the crud on the Ottoman army's boot. Moving out across the plain, Stepan's unit trudged past the town limits and the groves of twisted vines dotted with fruit. With most of the men away at war, this beautiful vineyard would most likely rot, as farming had plowed to a halt. After passing the rows of grapes, dangling heavy and close, the battalion marched toward a mountain in the distance that stood more than five thousand feet high.

So many times in the summers of his past, Stepan had visited there. He could still envision the gypsies and their young boys festooned in long *fustanellas,* traditional pleated skirts, holding *dumbeleks,* tom-toms, and instruments that shook, jingled, and jangled. He could almost feel the giddiness of those days, the absence of worry.

At the foot of the mountain, the captain hollered for them to stop. A long bumpy road stretched before the men. Smoothing the rough parts of the path would be their job, the old man explained. Stepan picked up a tool and began to swing, and he did not cease until it was time for his lunch of dry bread. Then he was back at it until sundown, after which he and the others traversed the nearly three miles to where they would sleep. His platoon was hardly cut out for this kind of work. Supervised by an Armenian foreman, the group included Stepan the *emanetji,* Hagop the coffee-shop owner, and Mardiros the *yemeniji,* or slipper maker. Around this time, Stepan spotted his tall friend Tevon

Harutunian, who didn't seem quite himself. His heart-shaped face was drawn; his soulful eyes dull from a high fever. Until called to serve, the silversmith had been in his bed, growing weak from some mysterious ailment.

Most of August and September passed like this, the men smashing stones into gravel under the hot sun and then pressing on to another spot in Izmid. At the time, government leaders were receiving reports of Armenians trying to cross the Russian border to help the enemy. The officials ordered the traitors executed. A secret initiative to monitor Armenian leaders followed. Oblivious, Stepan continued to break rock after rock, feeling fortunate to have each day's rations.

Everyone scrambled to get enough food, and Constantinople had only enough wheat for two months. Long lines formed outside the city's bakeries as people chanted, "We don't care for victories. Give us bread!"

In the evenings, Stepan gathered with his workmates and ate his insipid porridge made of liquefied potato. *Tastes like water,* he thought, but it was sustenance. His friends were his real comfort. Many sat silent in the darkness, especially the ones separated from their wives and children, the lives they'd left behind, while others raised their voices in song.

Hearing familiar melodies in Turkish or Armenian sent Stepan's thoughts back to his family. At home, he would often accompany them with his accordion, the fingers of his one hand dancing upon the keys, the other contracting the bellows. Without instruments, the band of conscripts from across the empire sang anyway, ballads about their mothers, about wishing to kiss their hands, about leaving their beloved and dear brides behind.

> Mother, Mother! I was called up and taken away,
> I wasn't given a rifle, but was enlisted in the labor battalion . . .
> . . . Everybody's hope was to come back,
> Days, days, I go in such grievous days,
> I go, I go, I go as a soldier,
> I go to break stones.

As summer turned to fall, the war spread across more frontiers. The conflict was not ending lightning-quick, as the Germans had hoped. Instead, both sides dug trenches across Europe, anchoring battles to a near standstill. It was only a matter of time before the war arrived on Ottoman soil. By early November, it had officially come to the empire. The Ottoman's battle cruisers had steamed into the Black Sea and attacked Russian seaports. Several days after this provocation, Russia declared war on the Ottoman Empire, and the Ottomans promptly returned the favor, followed by Great Britain and France. The Great War had drawn the world, and Stepan, into its abyss.

Stepan's tour was not ending, despite his nightly dreams. It didn't matter how many rocks he broke — the road ahead, the one he had hoped would lead him back to his family, had disappeared into the unknown.

In the month prior, the CUP had repealed the capitulations, the privileges that allowed Europeans to live and work in the empire, while exempting them from Ottoman laws. With these contracts gone, foreign embassies were searched, establishments shuttered; only their orphanages, hospitals, and churches were allowed to remain open. The rhetoric against the combatant countries was growing louder, and mobs thumped through the capital chanting, "Death to Russia!"

By the middle of November, Sheikh ul-Islam, the influential religious leader of Ottoman Muslims, issued a ruling — a *fatwa* — urging fellow Muslims to wage a holy war against the Allied Christians. In Constantinople, the English and French citizens and diplomatic staff, longtime fixtures of the empire, flooded the train station, clamoring to flee.

All this strife made Stepan yearn for the touch of his *mayr*, mother, for the sound of her voice calling his name once again. About ten other privates in his unit, he noticed, had slipped away on furloughs to kiss the hands of their mothers, like the song said, or the lips of their wives. Stepan approached the major's miserable servant, Murad, and negotiated a deal: for several pounds, he, too, could temporarily disappear.

Reunited at home with his family, the courier jumped back into his old work routine beside his brother, grateful to be supporting the household again and appreciative, too, that Armenag had paid the *bedel* and been able to keep the business afloat. The brothers' commute to the marketplace, however, traced a strange new landscape, one weighted by war. With the young men away, the streets seemed unnaturally full of women and children and the old. In the Uzun Charshi bazaar, where Stepan had once hung up his flyers, women and teenagers manned the shops, often at a loss about what to do. The world's conflict touched all the people of Adabazar, regardless of ethnicity or creed. "Every day towards evening, I went downtown to the marketplace and scanned the papers that had arrived from Constantinople," recalled one resident, Garabed Hovhannesian. "War was written all over them. Whether Turk, Armenian, Greek or Jew, all faces were sad without exception." Still, many believed the turmoil would conclude within six months, certain that the Ottomans would be no match for the Allied superpowers. One man even pronounced the war would terminate in two weeks.

With travel prohibited, Stepan stayed closer to home than usual on his furlough. But every few weeks, he ventured out to meet Murad, renewing his permit for his leave of absence with another infusion of cash. That December, the once-distant war was tearing at the edges of the empire like a ravenous flame set to a map. Already the Mesopotamian town of Basra had been lost, and the Ottomans trudged deeper into the Caucasus mountains on Russia's eastern border as winter bore down.

This was the heart of the ancestral land of the Armenians, who lived on both sides of the border, the two countries now at war. Encouraged by some early success in halting a Russian advance, the Ottoman war minister, Enver Pasha, marched his ninety thousand Turkish soldiers through mounds of snow in temperatures that dipped to thirty degrees below zero to confront the enemy on the main road straddling the frontier. Another Ottoman corps would surprise them from behind, near the town of Sarïkamïsh, traversing narrow snow-covered roads and towering ridges rippling with ice.

The wintry weather itself became another enemy. As the Turkish forces marched into the hinterland, many succumbed to the elements. Some finally managed to reach Sarïkamïsh, only to be decimated by a large Russian force.

Moreover, without enough food or tents, untold numbers of Turks froze to death, starved, or died from spotted fever. By the time the soldiers retreated in January 1915, only twelve thousand had survived.

Instead of admitting their own ineptitude and failure, the Turkish leaders blamed the Armenians for the heavy casualties, claiming the Ottoman Armenians had crossed lines and joined the Russians, their fellow Orthodox Christians. While some Armenians did fight elsewhere, they represented an insignificant proportion of the Russian army. There was no wave of Armenian defection to Russia. But there was a new narrative that shifted culpability for a tragic military loss to a whole civilian population.

Soon cries of revenge against the Armenians trumpeted through the empire. "In Istanbul, the propaganda work necessary to justify an enormous crime was fully prepared: the Armenians had united with the enemy, revolution was about to break out in Istanbul," wrote a Turkish man of the time.

The military challenges continued on the empire's northwestern coast, where a four-and-a-half-mile strait between the Aegean Sea and the Sea of Marmara exploded to the top of the world's news. On February 19, 1915, an armada of French and English warships and cruisers made a surprise attack on the Dardanelles, beginning what would become known as the Gallipoli campaign. The Turks were already stumbling from a failed attack on the British-controlled Suez Canal just weeks earlier. The Allied Powers hoped to open up the strait so they could take Constantinople and hasten the war's end.

As the Gallipoli campaign battered on, anxiety pulsed through Constantinople. People scrambled to depart as Ottoman leaders prepared to move the capital, the archives, and the gold, fearing the city would fall to the Allies. But the Ottoman army was able to repel the naval assault. Though the greater battle for the strait continued, the victory against the feared superpowers bolstered the Turkish leaders' resolve. They could now turn their attention to a matter within their own borders — and excise the Armenians like a tumor.

In the spring of 1915, following more than six months of a personally expensive furlough, Stepan found an unwanted surprise waiting for him at the train station: miserable Murad. "Return immediately to the platoon," he commanded. "The government has uncovered the scheme and our superior has been replaced."

At once, Stepan departed to join his work gang at the construction site in Chubuklu village. He had no idea what the government had planned for them. The Armenians serving in the regular army were already being transferred into the labor battalions. "Armenian individuals are absolutely not to be employed in armed service, either in the mobile armies or in the mobile and permanently deployed gendarmerie [units], nor in service in the retinue or offices of the army headquarters," read the order from the war minister.

As companies like Stepan's grew in size, rumors gripped Adabazar's Armenian quarter. Reports circulated that the government had sent "sealed letters" to the heads of *vîlayets* and towns about their Armenian residents, accompanied by instructions to open the letters at a later date. No one knew the contents. Sadness turned to panic as the Armenians, still raw from the massacres of the past, feared for their future and for the remaining population, composed mostly of women, children, and the elderly. "The Turks were silent, the Jews and Greeks were unconcerned, but the Armenian soul was perturbed," Garabed Hovhannesian wrote. "The 'sealed letters' were not good omens. The Armenians understood that 'wicked things' were hovering above them."

While Stepan toiled in Chubuklu, other Armenian platoons were being escorted off well-traversed roads. Dispatches of this were making their way back to Henry Morgenthau, the American ambassador in Constantinople. "In almost all cases the procedure was the same," he wrote. "Here and there squads of 50 or 100 men would be taken, bound together in groups of four, and then marched out to a secluded spot a short distance from the village. Suddenly the sound of rifle shots would fill the air, and the Turkish soldiers who had acted as the escort would sullenly return to camp. Those sent to bury the bodies would find them almost invariably stark naked, for, as usual, the Turks had stolen all their clothes. In cases that came to my attention, the murderers had added a refinement to their victims' sufferings by compelling them to dig their graves before being shot."

Sometimes, the bodies of these slain men who had labored for the Ottoman war effort weren't buried at all; they filled the ruts of the roads, like construction debris. Still blithely unsuspecting, Stepan shoveled, deeper into his nightmare.

With these killings, the decimation of the Ottoman Armenian population had begun. Not with a thunderous roar from the Turkish majority but rather in secret, away from the population's gaze, in the ditches and the back roads of the Ottoman Empire.

That April, not long after Stepan returned to his platoon, Turkish police began bursting into homes in Adabazar, homes that belonged to esteemed Armenian lawyers and merchants and politicians. They confiscated books, papers, and writings, ransacking places while the owners watched helplessly.

Then the arrests rolled out: fifteen men — including the school principal, Stepan Stepanian, whose student had lost his *P* during the alphabet melee, and the prominent politician, Antranig Genjian, from the socialist Hnchag Armenian political party. "We didn't know where we were being taken," recalled Antranig in his memoir. "Grief and panic was over the entire city. People tried to get as much news as they could about those exiled. Some of them even tried to persuade themselves that nothing bad was happening."

Night had already begun to descend on the Armenian community. On April 24, 1915, police fanned out around Constantinople and rounded up the empire's most prominent Armenian leaders. Many had been asleep in their beds. Hands in cuffs, feet in slippers, more than two hundred bewildered men were dragged to Central Prison, that torturous hole Stepan knew all too well. It was a veritable who's who of Armenians all in one place: the poet Siamonto, author of an Alphabet

Anniversary book, priests, educators, doctors, assemblymen, senators, deputies, music virtuosos. "You, too?" repeatedly asked one newspaper editor every time he spotted yet another prominent face. Scared and anxious, no one knew what was happening until the next day, when their names were read aloud from a blacklist. With the escort of hundreds of military police, the men were boated over to Haydarpasha, the railway station. From there, trains took them eastward, across the city limits, past the town of Izmid, and onward, many never to be heard from again.

For the Armenians, April 24, 1915, was the beginning of the end. The arrested men of Adabazar soon vanished the same way. Exiled by train, the men arrived in a barren area outside the town of Konya in the country's interior. Besides one official and some guards, it lay uninhabited — yet it was steadily filling up. Another caravan of one thousand soon joined them after they had marched for thirty-one days, hundreds of miles from their homes in the eastern village of Zeitun, hunger and illness wasting their bodies. Then two hundred more arrived. The banished hailed from Eskishehir, Konya, Ankara. Sixteen more funneled in from Adabazar, including one of the wealthiest grocers, Hagop Semizian. The remaining men in Stepan's hometown agonized: *Who was next?*

The news even reached Stepan's battalion. Armenag sent word to Stepan, begging him to obtain a *bedel*. Again, Stepan insisted it wasn't an option, that he'd be safer in the service — both men unaware that no place held a safe harbor anymore.

People We Don't Mention

2007

OUR TURKISH DRIVER careened into an alleyway clogged with pedestrians. He wasn't slowing. In a few seconds, we would plow into this crowd and lead the evening news. Reflexively, I slammed my foot down as if to brake, just as he leaned on the horn and magically cleared the path. Unclenching the car handle, I tried to relax. An hour earlier, I'd landed in Istanbul, my mother's birthplace, seventy-seven years after she had fled with her parents and sister Alice for the United States. The transition from a quiet, darkened airplane to the full tilt of city life had left me a little jumpy, my nerves already frayed from my near-constant worry over beginning this trip.

Unlike the beach-bound tourists, I would be spending the next month along a landlocked route once riddled with Armenian internment camps, and then I would cross into Syria and do the same thing all over again. I was feeling anxious about everything—utterly consumed by genocide and my grandfather's story, even though I was in modern Turkey nearly a century later. The fear wasn't always rational, I knew, but I had no idea how I might be perceived. *Should I say I am Armenian? Should I tell people what I'm doing? Or does the government already know?* My fear had been spiraling. Thank God, my friend Suzy met me at the airport. She had moved here several months earlier for a foreign relations fellowship and was in the cab with me.

The car turned up a hill and climbed. Nearby, the Ottoman apartment buildings of Galata decayed down a slope. I knew the name, having read about my grandfather huffing through the neighborhood with his heavy packages. The area had since fallen into disrepair but was starting to gentrify with foreigners again. Coincidentally, Suzy was one of them, having rented an apartment in his old stomping grounds. The cabby slammed to a stop, and we lurched forward. An ambulance was blocking the narrow cobblestoned street, its lights flashing. The driver muttered something in Turkish. "Dawn, we have to get out and walk," Suzy said.

Dragging my luggage down the near-vertical hill, past balconies strung with laundry, we glimpsed the Golden Horn, an estuary that bisected the city, streaking it with a magnificent blue. I was surprised by the winsomeness of a city with such a dark past. A busy roadway extended over the water. "That's the Galata Bridge," Suzy said, pointing. On the other side sat Sultanahmet, or old Stamboul, where my grandfather had often stayed in a boarding house, the hill spired by the six minarets of the Blue Mosque and the golden dome of Aya Sofya. I looked at the Galata Bridge, the sides crammed with fishermen and their dangling lines, and remembered it as the setting for one of my grandfather's practical jokes in 1910. "Boys, I will cross the bridge without paying the toll," teased my grandfather, then twenty-four years old. "If you succeed, the beer is on us," his friends replied, calling his bluff. At the entrance, my grandfather shut his eyes, swung a stick around, and made his way across for free by pretending he was blind; the complimentary beers afterward flowed endlessly into the night.

Post-nap, and a little less bleary-eyed, I set off to explore with Suzy. Just a few steps uphill, and already I was panting in the heat. "I think that's the street your grandfather used to take," Suzy said. Before leaving her apartment, we'd consulted my grandfather's journals and wrote down the names of places he'd delivered his packages. Just for fun, we started down the nearby road once graded with more than one hundred steps, then climbed back up again, passing tiny electronics shops and markets. Vendors pushed past us, gripping wooden two-wheeled carts piled high with red roses and vegetables. *Just like him,* I thought. How difficult life was, how difficult life remained. Eventually, we flowed onto a wide promenade, İstiklal Jaddesi, once known as the Grande Rue de Pera. In my grandfather's time, Pera was the city's European neighborhood, dotted with the foreign embassies and businesses. This was where Stepan had come during the Balkan war to telegram his family that Armenag had survived. On both sides of us now, in Beyoghlu, travelers and locals packed the outdoor mall, visiting the bookstores, art galleries, and baklava bakeries, even the Starbucks. Euro-disco thumped from ground-floor boutiques. Boys in jeans and T-shirts ambled alongside girls and their modern twists on Islamic modesty: form-fitting tank tops layered over long-sleeved shirts, miniskirts over leggings, scarves draped over their hair.

"That's a building made by the people we don't mention," Suzy said in a mock-conspiratorial tone. The Mïsïr Apartment rose up, balconies in front of windows, rounded in the center. It now housed galleries and a stylish restaurant with a 360-degree view of the city. Considered one of Istanbul's most spectacular art nouveau buildings, it was designed by Hovsep Aznavurian, a prominent prewar Armenian architect. Later, we passed a derelict edifice. "That's another building built by the people we don't mention. Ha, that's what I am going to call the Armenians," Suzy said. "The people we don't mention."

That night, the hours crawled to dawn. As I lay sleepless on an old futon, a beautiful male voice seemed to rise and fall over the city's hills: the muezzin's call to prayer. I thought about the surreality of it all, how my grandfather had combed these very streets, heard these same sounds, how I was already walking in his footsteps.

The air felt stickier that morning when I left to meet my translator. Slender and dark-haired, he was sitting on a bench under the Galata Tower, a cone-shaped monument that has been guarding this hilltop for more than six hundred and fifty years. Ethnically Armenian, Baykar was fluent in several languages and extremely detail-oriented, having planned every imaginable option for the Anatolia part of my four-week journey, from the food to the hotels to the train timetables.

But first things first: I needed a cell phone. It wasn't 1910 anymore, after all. As Baykar and I walked to purchase one, we happened upon the hospital where my mother was born, the building obscured behind a wall. Typical fortifications, I learned, for the "people we don't mention" and the Christian churches. It made me think about my mother and her fear of Turkey, her pleas for me to stay anonymous. "Do you tell people you are Armenian?" I asked Baykar.

"I always do," he replied.

Down the streets, Baykar and I walked, the old world pressed upon the new. "What is it like to be Armenian here?" I asked.

"My thesis adviser asked me if I had any problems here. Before Hrant Dink, we say no. After Hrant Dink, I can't say no." Dink was the Armenian newspaper editor assassinated earlier that year. His death sparked a rare demonstration of unity of Turks and Armenians. Thousands protested, marching down streets and holding signs that read *We Are All Armenians. We Are All Hrant Dink.*

Later on, I wandered alone, passing the spot where Hrant Dink died. It was just an ordinary sidewalk, pedestrians rushing along during a normal workday. It reminded me of the time I witnessed a cyclist killed in Manhattan and how an hour later, the normal city life wheeled through, as if nothing had happened, the street scrubbed clean of blood. This was how Turkey felt to me: the Armenians had been washed away, and the pace of life continued, the past eliminated.

It had been only two days since my arrival, yet it seemed like much longer. Somehow I felt as if I'd always been here; the stories of this place had wrapped around my life. The food sold on the streets, the *boereg,* cheese-filled phyllo, and the *lahmajun,* thin dough topped with minced meat, were the steak and potatoes of my childhood, happy alternatives to my mother's California-inspired grains and greens. The cavernous bazaars were filled with the kind of slippers I had had as a kid, decorative and curled at the toe, like Aladdin's. Most familiar of all were the people, olive-skinned, with thick eyebrows, so reminiscent of my Armenian relatives.

Just before leaving the next morning, I received an e-mail from my parents about Galata, "Our Dear Dawn, How exciting to live one block away from where Baba transported his heavy load up that steep hill. Since you are in the area, you should walk up that Hill, and congratulate Baba on his amazing stamina. To get the real experience, you should load your suitcase on your back and carry it up to the top. Love, Mom, Dad and the cat."

Though I didn't have the time — nor the stamina — to load a suitcase on my back and take it to the top, I did arrive at Haydarpasha railway station feeling a little more buoyant. Together, Baykar and I boarded the luxury train that would take us to our first stop — Adapazarï, the modern name of Adabazar. After stashing away my fifty-pound North Face rucksack, I began to think about all that lay ahead. As the car pulled out, I remembered my grandfather's advice, scribbled in a notebook, "Men go through life like walking on an unknown path. Every step must be taken in the dark." Closing my eyes, I leaned back and found my strength.

Following Orders

1915

FROM THE WORK SITE, Stepan could see masses of families streaming down the road under an armed escort. All of them were Armenians from the war-torn Dardanelles region, moving toward Chubuklu, their bundled belongings in tow. Stepan wasn't concerned about his own family until he heard about the expulsions in Sabanja, the lakeside village beside Adabazar. That July, the place had been emptied of its Armenian population. To Stepan, the upheaval didn't make any sense. The move could, perhaps, be justified for those living close to the sea, near the front lines and the enemy they were supposedly in league with. But why deport the people of Sabanja?

Anything could happen now. Panic seized him as his thoughts ran to his family. He wanted to sprint home, protect them, but there he was, trapped by the ditch with his battalion. Stepan and his entire company, nearly all from Adabazar, grew despondent. Beneath the season's merciless sun, they tried to labor on as more news reached them, this time about the nearby town of Bardizag. The details were slim, but something terrible had apparently happened to the men there. That night, the soldiers' gruel sat before them, cold and untouched, the day's closing bringing no sleep, no relief.

All their worrying, however, could not stop the events already set in motion at home. Some fifteen thousand troops had stomped into

Adabazar and the Armenian quarter, ostensibly to keep the peace. Along the main wide avenue, the brand-new Getronagan School operated as a hospital. The students had been moved to another location, unaware that they would be the school's last graduating class. Some soldiers billeted in homes and conducted military exercises while watching the residents' every move and meting out punishments for the smallest of infractions. It was like an occupation, only by their own countrymen. The summer took a turn when the cunning İbrahim Bey arrived and took command of the regiments. This was the same man who had pummeled Stepan, the former warden and terrorizer of the Central Prison. The populace gasped in alarm when they saw him; his reputation had preceded him. For weeks, all was calm, and the concern seemed unwarranted. Then came the July day when İbrahim's minions surprised the Armenian leaders in the bazaars, circled them with chains, and took them to the Surp Garabed Church, turning the sanctuary into the town's newest penitentiary.

At the front of the room filled with religious iconography, İbrahim Bey stared icily at the frightened men, crammed together on the floor. Lowering himself into a chair, flanked by a police chief and a general, İbrahim Bey began.

"Let me first introduce myself," he said. "I have conducted raids in the mountains and spent a lifetime in such activities. I can identify your revolutionary committees, whether Tashnag, Hnchag, or others. Many of their bosses were my old friends in the past and I've worked with them when our interests were the same. Now they've committed treason against the government, and as a consequence we oppose them. I've been personally here before with the Hnchag to distribute arms and know those who have weapons. At the time, bombs were brought in, too. Now I am here and commissioned to uncover those arms and bombs."

The detainees listened to his speech. While some belonged to one of the two political parties, others were clergy, no doubt already calling on the Almighty.

"If you surrender them, no harm will befall you," İbrahim Bey continued. "But if you hide them, know well that I'll bring upon you all types of calamities. You surely have heard what I've done in Bardizag. I filled in the church there with three hundred men and women and beat them, tortured them, made them lie on the floor and broke their bones within an inch of their life until they revealed where the arms were."

The *mukhtar* Artin Chalekian, the quarter's mayor and negotiator, asked to speak. Permission was granted. "These arms were brought in by the [revolutionary] Committees," he said, the ones previously set up by the political parties. "The Committees are dissolved now, and the townspeople don't know where the arms are hidden."

The *mukhtar* was referring to the weapons given to the Armenian armed units during the 1909 countercoup. After a movement grew to restore Sultan Abdul Hamid II to full power, many Armenians united to protect their CUP leaders, and in the southern town of Adana, the tension peaked. There, the sultan's supporters massacred some twenty-five thousand Armenians, punishing them for exercising their new constitutional rights. Fearing the same could happen in Adabazar, a secret political committee had held on to the arms for protection, just in case.

"They aren't dissolved," snapped İbrahim Bey, angered by the *mukhtar*'s remarks. "The arms must be brought out, otherwise all of you will go to Mosul and Deir Zor."

İbrahim Bey rose swiftly from his chair and exited the room; the air seemed to go with him, the men's breath gone. Mosul was a *vilayet* on the empire's eastern edge. The unfamiliar town of Deir Zor was closer, but still almost nine hundred miles away, in the Mesopotamian desert. Already, Armenians were being driven there, overwhelming the locals. But now Talaat Pasha, the minister of interior, had added a new rule: Armenians could not exceed "ten percent of the Muslim population." Where would they go? The answer was not clear. Beyond the settlements, there was only a vast, inhospitable desert.

Inside the church, one of the captors called Mukhtar Chalekian's name and gestured for him to come forward. He stood up and was escorted out. On his return, the captive men gasped at his bloodied and bruised body, stamped with İbrahim Bey's rage. The monster reappeared now, swinging his wooden cudgel around, bashing the men, all the while calling for the handover of weapons. The guards joined in, commencing the *falaka* or bastinado — a torture Stepan was horribly familiar with. Under the gaze of religious effigies, several men lay prostrate as waterlogged canes buffeted their tender soles until they lost consciousness. The rest, splayed across the floors, watched with dread, knowing they'd be next. Sometimes in these circumstances, the feet burst open, requiring amputation later. İbrahim Bey paid special attention to the Reverend Mikayel Yeramian, the prelate's second in command and father of Stepan's good friend. For hours the minister wailed, his cries echoing through the chamber. The screams were so loud, the sound traveled for almost a mile, over the rooftops and down the line of clustered homes. By the next day, İbrahim Bey summoned the rest of the town's men to Surp Garabed for a warning. With most of the strong males like Stepan conscripted or banished to the interior, the remaining men consisted of the elderly and those rich enough to pay the *bedel* — probably including Armenag. On that July morning, İbrahim Bey's audience increased to six hundred — all unequipped for what came next.

Outside Surp Garabed's walls, the streets were clamorous with shouts: "They are beating the men in the church to death, and they are going to begin on the women next!" "They say they will throw us into the River Sakarya." "They will send us all into exile."

Days passed, and fear descended. Armenian-owned stores stood shuttered, and only a few pedestrians shuffled about. Stepan's mother and sisters, like the other residents, stayed inside, hungry but forbidden to shop at the store or pick vegetables from their own gardens.

Near the church, a group of women kept vigil in a feeble attempt to save their husbands or fathers or at least slip them food. When one woman stepped in front of her disabled son to protect him, she was beaten too. The stolid guards wouldn't waver, even when a German woman tenderly begged for her betrothed. "Get out of the way or I'll beat you," the watchman warned. "I do not care for the German emperor myself, my orders come from Talaat Bey," the minister of the interior.

Each morning, two carts rolled up, both piled high with water-soaked canes, ready for use. With the torture instruments replenished, more men writhed and moaned from the beatings. As the assault continued, one man dared to talk back, "You must answer to God in Heaven for these things."

To which his tormentor decreed, "You have no God but me."

A few Armenians cracked under the pressure and volunteered to retrieve weapons. As they left, an Armenian stood at the door, marking off their names. Some purchased guns from Turks to appease their captors; others divulged the names of those with knowledge of this supposed cache. One priest snitched on an exiled political leader. Summoned back, the accused man stood before the church's jailers shortly after and asked, "Why do you punish these men? If there is any fault it is mine, and yet I also am guiltless. This society was organized with the permission of the government. You allowed us to obtain firearms."

In secrecy, some Armenians had hidden these weapons. One pile was behind residential houses, another in the Surp Garabed Church, and there were even more in a large trunk that had then been covered in tar and lowered to the deepest point of a well. Only four people knew of the locations. These men included the printer of Stepan's flyers, Harutiun Atanasian, and his comrade Garabed Hovhannesian, both Tashnag.

In the pressure cooker of the church, Harutiun stepped forward. With a high forehead and a penchant for dapper Italian hats, the printer had a commanding presence in front of crowds. A friend turned informer, he pointed out his own comrade, Garabed. Learning this, everyone was stunned by Harutiun's betrayal. Did he think he and his wife, Lusi, would be saved? Of the hundreds there, the worst punishment was reserved for a precious few, including Garabed. These men, in addition to receiving twenty-four more rounds of *falaka*, had their armpits seared with hot eggs, their teeth sawed with an adze, and their feet tenderized like raw meat. A week passed, and many of the detainees wished for death. A first cousin of Stepan's was one. He hurled himself out a window, fell, and slammed into a balcony, severely injured but alive. The stench around the men grew intolerable, a holy place turned torture chamber and unfettered bathroom.

Finally, ten days after the Armenians' jailing, the oppressors decided there were no more guns, and they released the captives. The men were ecstatic; their nightmare was finally over. All left — except a dozen or so who were paraded through town in shackles. "They did this to show us to the Turkish people as traitors, and troublemakers," recalled one of the men, Antranig Genjian. "[We] were taken in front of the city hall, to be pictured with bombs and arms. This picture was shown in all the public places, and published in Turkish and in French."

Not long after, two of these men, Khoren Yeramian and Krikor Kayian, were hanged in a public square in Constantinople, their rag-doll bodies an example to all.

In early August, the chimes of the church rang out and the faithful turned toward the sound. Before the assembled parishioners, the priest shared the devastating government decree. "Dear people, I have very bad news that I have to tell you," he said. "I'd rather fall here and die than to tell you this news. They have given us twenty-four hours to leave everything, our homes and everything, and whatever clothes we have on us and the money we have."

On her way back from the chapel, Stepan's sister Zaruhi broke into tears. When she entered her home, her round face betrayed the horrible truth. Her young daughter, Ovsanna, had never seen her like this before.

"Why are you crying?" asked the shy seven-year-old, her hair peren-nially worn in two long braids.

Zaruhi explained the deportation in a way that didn't worry her daughter. But no one — neither adult nor child — could truly fathom what was in store. Soon enough, the newspapers printed details of the "Temporary Law of Deportation" recently passed by the government's Central Committee that was engineered to steer the Armenians like cattle onto trains and roads to their end. "Army, independent corps and divisional commanders may send the population of villages or towns, individually or en masse, to other places and to settle them there in ac-cordance with military necessity and/or if they feel that they are spying or are indulging in treachery," it read.

Soon, signs hung in the quarters with instructions on departure, one district per day. Zaruhi discovered her neighborhood would be one of the first to leave; the rest of the family, who lived in another quarter, would be last.

"[My aunt] Arshaluys suggested that everyone in the family gather at her house so that we could all leave together," recalled Ovsanna.

With the departure looming, Armenag and each of his sisters looked around their homes anew, every blemish on their furniture and scrape on their walls telling a story of their lives. They sorted through all their possessions, stowed their most essential belongings in bags, and hid the rest — the beds, the pots and pans, the other household items — in secret compartments below closet floorboards, hoping everything would be safe until their return. They tried hawking small items outside, where a few Turkish residents looked but did not buy. "They knew that after the Armenians left, everything would be free," Ovsanna said.

Up and down the streets of Adabazar, families similarly prepared for the long journey. Some Armenians finished packing in just a few hours so they could arrive first in their new destination and establish their businesses in the promised flourishing city before the rush. Others, like the Miskjians, dragged everything on the sidewalk hoping to sell it, the pavement bedecked with valuables — Oriental carpets, chests, clothes, and jewels — like absurd booty from a treasure chest. One woman traded fifty rugs, worth fifty gold liras, for a pittance, as more furniture, china, and dowries were carted outside. Their lives were on display, split open like wounds. Some tried to conceal their money, hiding coins in belts. An American missionary stationed there watched as the desperate tried to raise funds for the journey. "All was very quiet — the silence of despair," wrote Sophie Holt. "Even the Turks looked serious, for they knew that their city was financially ruined, as the Armenians are the most thrifty and skillful of all the peoples of Turkey. They were told to leave their possessions in the churches and they would be safeguarded, but the same promise had been made in [Sabanja], and the church had been looted almost before the people were out of the city, so nobody trusted this promise."

That summer of 1915, a tempest swirled over Asia Minor. The political winds blew the Armenians into the deserts of the interior or Mesopotamia. The Ottoman gendarmes swept the minority Christians in the east toward the south, and the ones in western towns like Adabazar eastward into less populous regions — away from the capital and the world's gaze. Though Protestant and Catholic Armenians were spared at first in some places because of German pressure, Greeks were deported from the coastal areas around the Sea of Marmara. Elsewhere, Christians converted to Islam in the hope of avoiding persecution.

Despite the Ottoman government's censure of evidence documenting deportation, eyewitness reports nevertheless rolled into the American ambassador's office. Just two years into his post, Henry Morgenthau was growing increasingly alarmed. On July 10, 1915, he had telegrammed the United States' secretary of state, "Persecution of Armenians assuming unprecedented proportions. Reports from widely scattered districts indicate systematic attempt to uproot peaceful Armenian populations and through arbitrary arrests, terrible tortures, wholesale expulsions and deportations from one end of the Empire to the other accompanied by frequent instances of rape, pillage, and murder, turning into massacre, to bring destruction and destitution on them." Turkish authorities had told him not to interfere. Unsure how to proceed, he sought advice from Washington.

Days passed and Morgenthau didn't receive a response. True, it was a different communiqué than usual, and it concerned interceding in the affairs of a country not at war with the United States and involved matters irrelevant to American issues. Still, he sent another telegram on July 16, 1915, "Have you received my 841? Deportation of and excesses against peaceful Armenians is increasing and from harrowing reports of eye witnesses it appears that a campaign of race extermination is in progress under a pretext of reprisal against rebellion."

When the midnight moon shimmered overhead, Stepan knew it was time. He needed to see his family at any cost. Slipping away from his garrison, he joined ninety other Armenian laborers on the verge of es cape. In the semidarkness, the men set out toward the village of Armash, their eyes adjusting to the charcoals and blacks, the night yawning open like the road. After a few hours, the blush tones of dawn dusted the sky, revealing mountains. Not far from home, they knew. At the ascent, however, stood two guards on horseback. The band of deserters stopped.

The sentries sat up high, their weapons still mounted. *Not drawn,* Stepan calmly noted, perhaps a good sign. A guard broke the silence, and the warmth of his voice took Stepan aback. The sentry understood why they'd escaped, he explained, in these extraordinary times. "Return to your company, and you won't be punished," he finished.

The man seemed sincere. Still, the group needed a plan, and they huddled together in a frenzied exchange. "Resist instead of surrendering," implored the only five Armenians with weapons.

The remaining men — eighty-five of them, Stepan included — disagreed, and they turned toward the soldiers. "We agree to go peacefully," one of them announced, as if their acquiescence were a bargaining chip. Under guard, toward Armash they marched, where the Armenian seminary had already been emptied of its clergy. The crunch of the Armenians' shoes was almost muted by Stepan's chatter with the guards. *They're friendly,* he thought. *There's nothing to worry about.* He believed this, even though the guards stood at spaced-out positions, ready to fire if necessary, on the route back.

Finally, the escort entered the public square of Armash, where the mayor promptly admonished them for their "boyish exploits" before calming their nerves. "You aren't going to be punished. You'll be safely returned to your posts." For a moment, Stepan felt comforted. Until he saw the gendarmes circle them. Until he felt his hands pulled behind his back and tied. After escorting the Armenians to a government building, the sentries locked them in the basement. Stepan was wedged between the others, upright all night, limbs jammed against limbs.

Back home, the Miskjians were almost ready. What food they had for the journey was cooked and wrapped; their bags were nearly packed. They couldn't take anything that heavy; no one knew how long they'd be gone or the distance they'd have to walk. As they dashed around the rooms, a note arrived, delivered by their neighbor Boghos, who had escaped from Stepan's battalion. The paper bore Stepan's distinctive handwriting, that small, neat script. A long letter, it closed with urgent instructions, "Make absolutely certain to take along my valuable possessions. You can sell them for food. I wish you a safe journey and remember that this could be our final separation — not to meet again. Sending kisses to all of you."

It was a son's farewell to his family. Every moment now seemed to be the last one, and not seeing Stepan again burned their insides like acid. All this was too much for his mother — the deportation, the loss of her home, the loss of her younger son. Hripsime broke into tears, as did her children, all weeping for him and perhaps for themselves too. But there was no time for hysterics for the last quarter was departing. And there could be no delay. At another house, the guards kicked out the lingering family members and confiscated the keys. The grief-crazed mother stood frozen outside the door, wailing. She pressed her lips to the hard surface, kissing the house her family had constructed. The owner of one residence simply left his door unlocked and hit the road, explaining to his son that it was useless to do otherwise; the Turks would break it down. At yet another home on these sorrowful streets, a grandmother summoned her grandchildren to her. "God listens to children's prayers," she explained. As they surrounded her, lifting up their small arms, she said, "Make our journey safe, and soon make us come back and find our home as we left."

Some mothers were so distraught, they tried to hand their babies to a local missionary. But a Turkish officer intervened. "These Armenians are dangerous people," he said. "They may have bombs."

Bags in hand, Stepan's mother, siblings, nieces, nephew, and brothers-in-law crossed under the door's archway and strode down the three stone steps as they had so many times before. Aghavni had joined them with her three young children, Araxi, Madeline, and a baby, Louise; Zaruhi was there with hers. Thirteen people — three generations — all not knowing if they would ever ascend them again. Everything they and their neighbors had worked for in this town — the assembly hall, the schools, the churches, the gardens — were left behind as they pushed toward Adabazar's border.

Pedestrians choked the streets, and trucks barreled along with seventy people crammed inside, almost twice the capacity. They were all heading to the station in the town of Arifiye, following a current more powerful than the Sakarya River and its hurricane of floods. Like many Armenians, the Miskjians carried their belongings in *bohchas,* square hand-stitched swaths of cloth with sticks as handles. The four corners were tied into a bow, the heavy burden slung over a shoulder or back. In the past, when transporting a load, Stepan had taken a cart to cover the five-mile distance.

Of course, these weren't normal times, and they were lucky to have a ticket. These rides were not free, even though the government had mandated the deportation. A wealthy man had paid for the Miskjians and innumerable others. Without the benefit of such generosity, the less fortunate began their long walk toward the desiccated interior.

Even with all the trains coming and going, thousands of people had already been bottlenecked at the station because the cars were unable to remove them fast enough. Some had slept there the night before. The railroad had promised to bring progress, not this. With open eyes, the Miskjians stared down the tracks and waited for a train to take them into the unknown, uncertain if they'd ever return to the divine city, previously under God's care.

Under the Black Tree

2007

AS I STEPPED OFF THE AIR-CONDITIONED TRAIN IN ARIFIYE, the heat blasted me. August was always like that, I was told, and that day, the temperature was expected to climb to ninety-eight degrees. My translator, Baykar, and I hailed a taxi and were driven onto a large, paved thoroughfare. Beside us, long stalks of corn swayed, and thickets of trees reached for the sky. Every few minutes, a boxy modern building interrupted the view. The driver accelerated, veering in the direction of a blue sign that read *Adapazarı*.

For as long as I could remember, I had heard from my mother about the wonders of this place: the decorative Ottoman houses and narrow bazaars, the beautiful water mill that spun like a Ferris wheel. She had never seen the rolling paddles herself; she had only heard about them. It was a town my mother always told me was mine. She had passed this inheritance down to me, along with her and her parents' olive complexion and deep-set brown eyes.

In the back seat, I unzipped my pack and took out a sheet of paper, a copy of an old map from a book, the curves of avenues and intersections drawn by hand. Squares depicted the churches and the schools, the Armenian cemetery and taverns, all the important sites — at least, important to the Armenian who had rendered it. A Muslim Turk would have charted the six mosques; a Christian Greek the two Orthodox churches. My grandfather's neighborhood, Nemcheler, was in the top right corner. With a magnifying glass, my mother and Arlene had translated the Armenian script labeling the landmarks. In cartoon bubbles, my mother had written the English words in all caps: MASSIS THEATER, MANOL'S SEA, WATER FOUNTAIN.

I glanced up and saw a sprawling city. "Are we here?"

"Yes," the cabby answered in accented English. Computer stores gleamed near kebab stands, traffic islands, bright signs, and commercial edifices. It was nothing like the black-and-white photos I'd seen that depicted a simple Adabazar, one I had imagined still existed, at least in some form. Many parts of Istanbul were similar to their images on early postcards, but this city, so far, was completely foreign. I began to worry about what I'd find here, if the past had been irrevocably lost. Nearby, trees, planted in a line and saluting the cars, were an obvious stab at beautifying the unbeautiful, a municipality that had grown large and been rebuilt without care. Ninety-two years ago, about half of Adabazar's population of around thirty thousand people were Christian Armenians, with some Jews and Greeks. Now it was predominantly Muslim. The driver circumnavigated a large man-made park before pulling up to a spartan hotel with a carpet that spilled onto the sidewalk, welcoming guests. After checking in to our bare rooms, Baykar and I met downstairs.

"Do you speak English?" I asked the young desk clerk.

He nodded.

"Do you have a map?" I asked. Baykar hovered near me in case he needed to translate.

"No."

"I want to walk around and take in the sights," I explained. "Where are the Armenian churches?" I couldn't remember the last time I'd visited a church, but I needed to see one here. In my grandfather's time, there were four Gregorian ones — Surp Hreshdagabed, Surp Garabed, Lusavorich, and Surp Stepannos — spaced almost equidistant from one another. If I could find them, I told myself, I would know where to go.

The clerk's eyebrows lifted at this strange request, and he shook his head. Though it might have been obvious, I refrained from telling him about my background and focused on learning as much as possible. My map was useless so far; not even the expansive park was depicted.

"How about the hamam?"

"Earthquake destroyed," the clerk replied in a heavy accent about the bathhouse. Same with the old part of town. The temblors shook so frequently that Adapazari had its own museum devoted to the subject; exhibits documented the calamitous years with Richter scale measurements: 1943 and its 6.6 quake, 1967 and its 6.8, and 1999, when a 7.4 rattler destroyed twenty-six thousand buildings and claimed nearly four thousand lives. I had read about the 1999 disaster in the newspaper but hadn't fathomed how much of this town had been destroyed. Even the water wheel where my grandfather had spent so much time was gone.

I moved on to something I thought an earthquake could not ruin. "In what direction is the Sakarya River?" I asked. "To the left," he replied.

"How far?"

"Can't walk."

"I thought it was close by."

"It used to be," he said. "It moved." Whether this was true or not, I didn't know; the clerk seemed impatient now, bewildered by my barrage of questions. Glancing around the empty lobby, I began to panic a little. I'd come all this way; had it been a waste? What if I didn't find anything?

It seemed like I'd lost all my markers and wandering around later that afternoon didn't really help, as we passed clothing stores and bakeries but nothing from my grandfather's time. Finally, I went to see if the Turkish cemetery on my map still stood. Sure enough, on the outskirts of town, I found one, the long raised graves shaded by trees. As we were winding our way back, I kept guessing the location of the old Armenian neighborhood, and after enough twists and turns we came to Karaaghach Boulevard, once the heart of Under the Black Tree neighborhood. I searched for Armenian writing, for the Massis Theater — anything — but couldn't find the roots of the vibrant community anywhere, only the shadow the black tree had cast down on my family for generations, the shadow of loss.

Where was my family's Adapazari? I thought about the old Adaba-zartsi women who used to come to my house in Los Angeles. As a child, I hadn't realized the significance of those reunions with my mother. Each time, it was the same. My mother and I would position all the metal folding chairs in front of the mantel and the arched stained-glass windows. I would be so excited, since the gathering meant I could pre-tend to be a ballerina and spin around in front of the trapped audience, despite having no talent at all. After the doorbell rang, the plump la-dies would stream in carrying dishes covered in aluminum foil. Many came dressed in black, their hair in buns, a few wiry strands sprouting from chins. These Adabazartsi natives had found their way to our liv-ing room, half a world away from their birthplace. Many were widowed. One of them had lost her first husband at the outset of the deporta-tion, when he was blamed for building bombs and hanged; her children would one day bury her with the love letters he sent her from jail while awaiting execution. But on these days, the ladies tried to focus on the good times, and for one tiny moment the town seemed to come alive again. They could still taste the sweetness of the local fruit on their lips and the alcohol at Manol's Sea, where Stepan's friend Harutiun used to get drunk. They could remember the faces of loved ones, lost but never forgotten.

My mother was not a natural cook, but she worked hard in prepar-ing the feast for our visitors. After the death of her mother, Arshaluys, who also hailed from Adabazar, she vowed to continue her tradition of hosting the expats. In our avocado-green kitchen, my mother would suppress her vegetarian instincts and make lamb with bulgur, which would be surrounded by the delicacies that the women brought — *pak-lava, sarma, köfta,* and *tutumov boereg,* cheese squares stuffed with pump-kin. She couldn't quite outdo her own *mayr,* whose parties were not to be missed. "For four days she cooked," she said about my grandmother, a large woman with few wrinkles, even in old age. "She'd spread out a plain white sheet on the queen bed, and then prepare sheets of the phyllo dough on top, just as large." She would mix up the stuffing for the cheese *boeregs* — the chopped parsley, the feta, and the butter — place it on strips of dough, and fold the strips into tiny triangles with the goodness inside. In waves, she'd bake them for the sixty or so corpulent guests.

On our tree-lined street in Los Angeles, the old women talked with my mother and ate and pinched my cheeks. *"Shad anush aghchig,"* what a sweet girl, they would say, smiling, thumb and index finger mimicking forceps, leaving a redness on my cheeks akin to a burn. Beyond those words, I didn't understand much; neither did my father, an American whose Scottish-English lineage traced back to the Declaration of Independence and President Grover Cleveland. Until adulthood, I found his history far more fascinating than my mother's chronicle of poverty and a peddler father. The upper-middle-class MacKeens were built into the foundation of this country, into everything I'd learned about in school. The Miskjians, by contrast, were poor immigrants from a country seemingly named after part of our Thanksgiving dinner.

Inching even closer to me, these ladies would say in stilted English, "You are an Adabazartsi."

"Yes, I am," I'd reply, somewhat reflexively.

Then my mother would grandly declare, "We are all Adabazartsis!," her arms open wide. She loves these kinds of pronouncements, both believing in this sentiment and wanting to make people laugh. It wasn't until I was an adult that I realized my mother identified herself as a native of a place she'd never been. To her, Adapazarï was more her origin than Istanbul (her birthplace), New York (where she had spent most of her adolescent years), or Los Angeles (where she married, bought a home, and had children, and where she has resided for nearly six decades and counting). With my stable upbringing, never moving houses, I couldn't understand her fidelity to this foreign place so far away.

The next day in Adapazarï, Baykar and I set out in a taxi to find the Sakarya River. As farmland rolled by outside, our bearded driver smiled into the rearview mirror, his eyes crinkling, the top of his traditional *jübbe*, a long white robe, visible above the seat. His warmth was contagious, and I returned the same wide grin. Still, I wasn't sure if I could trust him. The town had become very nationalistic and conservative, I'd heard, and I didn't know his views.

"It's there," the driver finally said, pointing to the right as Baykar translated. The car slowed and then halted.

I peered out the window. *Like a large puddle,* I thought. The river was low and muddy, nothing like the ferocious beast that used to regularly flood the town. We exited the vehicle and followed the driver onto the bridge, painted the electric blue and yellow of a tropical discotheque.

"There's a drought," the driver finally offered. "Usually it's much higher." Brown cows lounged on the other side, their hooves barely submerged in the low water. We stood there gazing out at the parched riverbed.

"Where are you from?" the driver asked me in Turkish.

"America," I said, weighing what would come next.

"Where?"

"Los Angeles."

Through Baykar, I steered the conversation back to the river. While chatting, I agonized over whether I should share the reason I'd come. I hadn't planned on it, but I didn't want to deceive him. Still, I could hear my mother's high-pitched voice: "Don't tell anyone why you are there! Don't forget what happened to Hrant Dink!" She'd repeated it constantly in the months before my departure, petrified of the place she had made me love.

"My grandparents are from here," I said carefully.

"Oh!" he said excitedly.

"But they weren't Turkish," I clarified. I glanced at Baykar; his expression assured me it was okay.

The cabby seemed confused.

"They were Armenian." I waited a moment, worried that his opinion of me would change.

Then his wrinkled face creased into more wrinkles, betraying a grin. *A green light,* I told myself.

"I want to find the Armenian cemetery, where my family was buried, and pay my respects. Do you know where that is?" Baykar patiently translated my request. On my map, it was in the top right corner, but it was hard to determine its true coordinates.

"No," he said. "Maybe we can ask."

He drove for a few moments and then pulled onto the shoulder. He hopped out and talked to several men milling in front of a farm stand. I felt nervous as they studied us, their faces stripped of emotion. Would they report me for coming, for trying to excavate the past? A few minutes later, he returned, my concern unwarranted; they were just trying to help. "They don't know. One said maybe farther down the road."

I had pictured finding the grave of my great-grandfather Hovhannes Miskjian, who had died when my grandfather was a child. Maybe I'd clear off the weeds, I thought, read the inscription, place some flowers there. Find a deeper sense of home than I had in the strip malls and freeways of California, a region where neither of my parents had grown up. We pressed on, into even more rural countryside. I tried to remain optimistic.

The driver slowed and pointed. "Maybe there," he said. I looked out, hopeful. There were no headstones, only vegetables; it was someone's farm. Perhaps we had the wrong place. We continued driving, but there was nothing for me here. Nothing tangible to hold, the imprint of my family and the rest of the Armenians seemingly erased. Soon after, we left Adapazarï, and I felt the loss all over again. I was buoyed by the driver's eagerness to help, but as we sped toward the interior, my previous anxiety proved to be well founded. My travels would soon attract the attention of the police.

Night Train

1915

BY THE TIME THE SUN ROSE, Stepan had been standing upright for hours, his legs pinned. In this basement, on all sides, stood the rest of the deserters, tightly packed in summertime sweat. *Pickled sardines,* he thought, *that's what we are.* How much longer would he be canned like this? Just then, the gendarmes came in and stood before the mass of ninety men. It was morning, time to get moving. They untangled the deserters and prodded them outside to relieve their bowels, a feat with restrained hands.

Nearby, some villagers congregated. All were women and girls, their men most likely drafted away. They held food and water, and he devoured his portion; the last time he'd eaten had been hours before. Soon, guards bound the captives' hands more securely and led them into the scorching sun, crossing one town's limits, and then another's, always trotting beside them on horses. Stepan's throat burned with each footfall. He'd never known thirst like this before; the gendarmes refused his pleas for water. Along the route, crowds of Turks gaped at the road show of men, dirty and panting like dogs. "Have mercy," the sympathetic ones beseeched the guards. They couldn't bear to stand by and do nothing, no matter what rumors they'd heard of the traitorous Armenians. Hauling jugs, the Turkish villagers would rush up, defying the guards' commands.

At last, Izmid's ragged coastline came into view, along with the dreaded military base. Stepan was about to face the tribunal that would decide whether he lived or died. He'd appeared there two years earlier, after the alphabet-anniversary debacle, and ended up in prison. He wondered what awaited him now.

The tribunal did not waste any time. Shortly after his arrival, Stepan was forced to the ground, face-down, his wrists untied. Swiftly, he felt the sentence in the soles of his feet: *falaka*. How quickly he remembered this from prison, how quickly the mind can forget the intensity of pain. As the wooden clubs thwacked down on him and the other escapees, some of the batons broke on impact; afterward, the men's feet often resembled cooked sausages, swollen and split. A gendarme ordered Stepan to rise. He stood on one foot, writhed, and then gingerly set the other down. To stand felt unbearable, yet that was the easy part. Through the night, the deserters were marched, hobbling the twenty miles like a band of invalids learning to walk.

By morning, the aching men reached Chubuklu, a village to the west of Adabazar where they'd witnessed the first deportations. At last, they could sit and rest. Back to breaking stones the next day, Stepan again plotted his escape. He could barely move, but that didn't stop his planning. Nothing did—until he learned something more agonizing than the bastinado that still stung his skin about his Adabazar. The Armenians there had been forced out, the people had said. It had happened maybe two weeks earlier. The heart of his beloved hometown pierced, stripped of life.

Stepan tried to picture his family gone. He tried to imagine the streets—and his house—empty. If it was true, then escaping to go home would be pointless.

As his company was returning to Izmid a short time later, they paused near the railroad. A train sat on the tracks that late August day, its passengers idling nearby. Stepan could see them, talk to them. "What is happening in Adabazar?" He had to be certain. Their reply left him crestfallen. "There is not a soul left."

Two weeks earlier, at the Arifiye station, the Miskjians had searched the sea of bags and faces, looking for a spot together. One of them cradled the infant Louise, Aghavni's daughter; another held the hands of her two small girls, Araxi and Madeline. It had been only a day or so since they'd been forced from their home; everyone was there, everyone but Stepan.

At eight, Ovsanna was old enough to register how strange this all seemed. "There were no seats," she said. "Just four walls." That left only the floor. They set their *bohchas* down and sat on top of them. All around squeezed the old and the young and the middle-aged; the temperature outside breathed summer hot, inside even more so.

The packed car bounced and rattled forward out of Arifiye station, just one train of many crisscrossing the Ottoman Empire with human cargo. The official proclamation in one town had outlined the conditions of travel, "To assure their comfort during this journey, *hans* [rooming houses] and suitable buildings have been prepared, and everything has been done for their safe arrival at their places of temporary residence, without their being subjected to any kind of attack or affronts," it read. Gendarmes would also ensure the Armenians' protection. "Where are we going?" many would remember asking. "To the interior," the gendarmes would answer. One rare photograph captures an image that would become iconic in the next war: three boxcars filled with hundreds of minorities, their dark eyes looking outward at a world standing by.

The Ottoman Armenians were being forced out of their homes and sent to places with unfamiliar names: Konya. Qatma. Dibsi. Deir Zor. Many had to walk, even the sickly. With Stepan still in his battalion, he didn't understand the reason just yet. He didn't know that this was not a "relocation"; it was a death march. And unlike Stepan's walk following his desertion, when he first experienced thirst, not all Turks felt sympathy. "Even veiled women threw large stones at us," recalled one Adabazartsi.

As the Miskjians hurtled toward the unknown that summer, Henry Morgenthau telegrammed the U.S. secretary of state again. "Turkish anti-Armenian activities continue unabated," he wrote. "Horrors to which large numbers of innocent and helpless people of this race are being subjected. Armenian population is fast being swept from Ada Bazaar and Ismit . . . It is difficult for me to restrain myself from doing something to stop this attempt to exterminate a race but I realize that I am here as Ambassador and must abide by the principles of non-interference with the internal affairs of another country." Stepan's birthplace had been nearly cleared of Armenians. In his county, 56,115 Armenians, 93 percent of the area's Armenian population, had been deported, and that was according to the figures of the interior minister.

Armenians kept appealing to their church's highest Gregorian leader, the patriarch Zaven Der Yeghiayan. His Holiness was in Constantinople and at a loss over what to do as he received reports on the destruction of his people. Writing to foreign embassies and countries, he begged for help. "The Armenians of Turkey are living [the] last days of their lives," the spiritual leader wrote. "We have no means for putting off death. If Armenians abroad are unable to move the conscience of the neutral states, very few of the one and a half million Armenians will be left within a few months. It is inevitable that they will perish."

Already, the Armenian leaders taken on April 24 had been massacred, and the residents from the Black Sea area had been ferried offshore and shot, their bodies dumped overboard. In the six eastern Armenian-heavy provinces of Erzerum, Van, Bitlis, Diyarbekir, Sivas, and Trebizond, entire villages had been yanked out of the soil like mature trees. For thousands of years, eastern Anatolia had been their homeland; it was where the majority of the Ottoman Armenians lived. Now their houses were filled with strangers, *muhajirs,* Muslims displaced from the Balkan Wars, and their wealth had been requisitioned.

In the northeastern region, the exiles were forced to march. Days of walking, goaded by the whip's lash, led to weeks. Alongside lines of five thousand, the gendarmes galloped on horses, guns slung at their sides. Just one guard for every three hundred Armenians. Numb and pained, they tramped forward, these prisoners in their own land.

A day or so after leaving Arifiye, the train carrying Stepan's family rolled into Eskishehir. Normally, it took six hours to travel the eleven stops southeast to this town, but the rails were overloaded with Armenians and troops, and the packed deportees languished onboard endlessly, without toilets or access to more food and water.

Outside, a large station towered several stories. In the sky's dimming light, Ovsanna looked around. At the time, masses of Armenians were loitering in the area beside the terminal. Beside crude tents — propped-up sticks with carpets as roofs — the people grilled scraps on open fires and hawked loaves of bread and fruit. It had the air of a festival, from afar.

A closer view revealed a dismal state. With the arrival of so many villagers, the station had been transformed into a transit camp. Officials at the U.S. consulate heard the transient population had swelled to twelve, perhaps fifteen, thousand. Away from home and hospitals and hygiene, the stranded Armenians were dying, the death toll as high as thirty to forty a day.

The interior minister explained to the American ambassador that August why he was taking such drastic measures with the Armenians: "In the first place, they have enriched themselves at the expense of the Turks. In the second place, they are determined to domineer over us and establish a separate state," Talaat Pasha said. "In the third place, they have openly encouraged our enemies."

Morgenthau countered his claims, detailing the exchange in his diary. "It is no use for you to argue," Talaat Pasha replied. "We have already disposed of three quarters of the Armenians; there are none at all left in Bitlis, Van, and Erzeroum. The hatred between the Turks and the Armenians is now so intense that we have got to finish with them. If we don't, they will plan their revenge."

While in Eskishehir, Ovsanna believed they'd stay. But that was not to be. Soon after arriving, the family was pushed onward. Their train moved across a plateau into the dry interior of the country, far from the blue waters of the Sea of Marmara and the silkworms that had once danced in the Miskjian family home.

Thirty miles past Afyon Karahisar, the Miskjian train stopped. The passengers were hustled off, the stations farther east too crowded to accommodate them. On the platform, they surveyed the scene. The station was called Chai, the backdrop unsettlingly gorgeous. Two and a half miles away lay the postcard-picturesque village of Chai, sloping down a mountain. Fifteen hundred houses were nestled together there, a stream of fresh water running through. But the Miskjians would not stay here. There were no *hans,* or "suitable buildings," waiting for them as promised in proclamations, no accommodations arranged by the government. Instead, a field stretched before them; this was their new home.

Despite censorship, news of the "race extermination" was beginning to reach the rest of the world. Pleas to save the Armenians arrived at the U.S. State Department. Morgenthau tried to persuade the Ottoman Empire's ally Germany to assist in stopping the persecution. The German ambassador did raise the subject, but the massacres continued nonetheless. Once again, the Turks warned other countries not to interfere in their domestic affairs. It was an effective defense when combined with foreigners' reluctance to intervene. The headlines in the United States and the international press told of the fallout: "Turks Are Evicting Native Christians," "Wholesale Massacres of Armenians by Turks," "Report Turks Shot Women and Children: Nine Thousand Armenians Massacred and Thrown into Tigris, Socialist Committee Hears." Following Morgenthau's appeals for aid, the American Committee for Syrian and Armenian Relief was formed, and it solicited money for bread and medicine. Still, on August 18, 1915, this headline appeared in the *New York Times:* "Armenians Are Sent to Perish in Desert; Turks Accused of Plan to Exterminate Whole Population — People of Karahissar Massacred."

Weeks after Stepan found out about Adabazar, his battalion was ordered to an area near Constantinople. For three days, he traveled by carriage to the small town of Beykoz, its base already crowded with two hundred Armenian laborers. Having recently been tapped to help his major, Stepan had been spared the march and felt proud of being singled out by his superior as "the most respectable man" of his division. That September, the major treated him well. For the first time in a long while, Stepan felt dignified, hopeful even, as he completed his administrative tasks. Perhaps the war would end soon, and he could return home. He thought this as the grounds mushroomed with more Armenian battalions, the number now six hundred, amassed there for some unknown reason. Two weeks later, the major received an order. He had to send two hundred workmen to Haydarpasha. Stepan watched them file into position and fade from view.

A week later, another telegram arrived containing the same directive. Two hundred more Armenians trailed away. Another week, another telegram. Yet more privates departed to Haydarpasha. Then the major summoned Stepan. A fourth telegram had come in, he explained. He had tried to refuse the order but had been overruled by his superiors. "Unfortunately, I can't keep you here," he apologized. All forty-seven remaining laborers, including servants, including Stepan, had to go. The major looked anguished, but Stepan couldn't yet fathom why.

Within a few days, Stepan was shipped out to a village close to Haydarpasha station. Innumerable labor companies were already assembled there, all the men about to move out, on the precipice of the Medz Yeghern, the Great Crime. There, Stepan saw his friend Khoren Akhbar Mkhjian, who was leaving before him. Back home, Khoren Akhbar — Brother Khoren — used to drive a cart and help transport the *emanetji's* parcels. "In case you meet my family, tell them I will be transferred within a few days," Stepan instructed him. "According to my latest information, they are somewhere next to Chai train station in Afyon Karahisar." He'd learned this by word of mouth, his main source of news. Chai was more than three hundred miles from Constantinople. What was the chance Brother Khoren would be able to find them in a mobbed field? Especially while under guard on a passing train? Stepan knew it was almost impossible, but that didn't stop him from hoping.

In the early-morning light of that October day, a train waited on the tracks for Stepan and his labor battalion. It was completely different than the usual passenger cars. *For animals,* he thought as he climbed into one, the iron bars sturdy across windows. All around, the laborers piled in until there was no space, and then more squeezed in, the boxcar corralling more than sixty head.

He could feel the forward momentum. Through the doors, swung half open for room, Stepan glimpsed the empire's capital. Beautiful Constantinople, with its seven hills, minarets, steeples, and boats that bobbed on cobalt waters. An unsettling thought came to him: *this will be my last image of the city.* "Farewell," he said, just under his breath.

As men's knees and shoulders jabbed him, the chain of cattle cars chugged eastward, shadowing the coast. Stepan looked about. To his left sat a guard, to his right sat another, both with old-fashioned rifles slung over their shoulders. They were the rulers of his new world. From their ragged clothes, he could tell they were peasants. From their unpolished demeanor, ruffians. Nothing more.

The morning stretched on like an elastic band, fitting in more moments than the hours could normally hold. The Sea of Marmara was chasing the servicemen along the compartment's side. *Such desolation,* Stepan thought as the color of life was erased. At each stop, he asked permission for his most basic need, which now had a price tag — to be allowed to urinate, he had to pay twenty to forty *paras,* double that amount for a bowel movement. Possessing only a small stash of money, he felt worried. He was literally pissing it away.

In the afternoon, lush mountains signified he was nearing home — not that Stepan could get off the train. He was a handful of miles from his house, his vacant house. Two months earlier, at this very terminal, Arifiye, his family — and his entire community — had departed for the interior. Perhaps Brother Khoren had located his family in Chai, delivered his message. Perhaps not. Still, Stepan rehearsed in his mind what he would say if given the chance to see them again. Each mile crossed, he was that much closer to them. The tracks rose and fell, passing the village of Bilejik, balancing between tall forested rises. All the while, Stepan couldn't extend his arms or legs. He felt the lice crawling from man to man. He was hungry, regretting not buying more provisions before they departed. The only food they could procure was from the guards, sold at exorbitant prices. At sundown, they stopped in Eskishehir, the day still retaining enough light for him to see. A field flanked the station, jammed with refugees. A mob came rushing toward the compartment's door, the faces searching for lost loved ones. The sentries trailed them, beat them back. This was the rhythm of the new world and the first refugee camp Stepan would see.

From Constantinople, the increasingly distressed American ambassador had tapped out another telegram to the secretary of state. "Destruction of Armenian race in Turkey is progressing rapidly, massacre reported at Angora and Broussa," he had written. "Raise funds and provide means to save some of the Armenians and assist the poorer ones to emigrate."

Stepan was still hours from Chai. His family had traveled this very same corridor, and now he wasn't far from them, the feeling intoxicating. The terrain grew flatter, then rippled along the side with rocks and hills. He was sleepy but wouldn't succumb, he told himself, no matter the pull. He had a small match handy; he'd light it and search the station's platform for his family. The train crawled so very slowly, and his eyelids rolled heavy, his head's weight like a boulder. More miles still, and the hills climbed into mountains.

Everyone Stepan loved — his mother, sisters, nieces, nephew, brothers-in-law, and brother — lived in two tents, positioned side by side. "We made a hole in the ground, then placed the wood in that hole," remembered Ovsanna. "[We put] a thick piece of cloth, sturdy enough to protect from the rain, over it."

Each day in camp felt indistinguishable from the next. The days it stormed, the days the sun shone bright, the thirteen tried to create some semblance of life amid the thousands of others struggling to do the same.

"I had no notion of time," Ovsanna said. "I had no school, no church. Nothing to do." No word on when they could return home; only the arrival of fall, the threat of winter. In this exiled life, no fences trapped them, and yet, Ovsanna recalled, "I don't remember trying to escape. Where could we go? There were Turks everywhere." A much smaller war was being waged in the tents, the battlefield often just a head of hair or a shred of clothes. Ovsanna remembers the enemy, little blood-suckers, and the endless biting, and her attempts to bleach the lice away.

Sometimes, the camp's din was broken by the calls of locals hauling food to sell. At other times, the area's men came for a different pur-pose: to carry off the prettiest of the girls. So far, only a few girls had been snatched, as the Armenian men had banded together to ward off the locals; they preferred death to losing another daughter. In other areas, the international news reports told of eighteen-year-olds being divvied up by looks, the best offered to officers, the less attractive to the infantry and the "highest bidders." Just young women, Stepan's sisters Mari and Arshaluys were afraid. Pretty, with taut skin, big brown eyes, and thick chestnut hair, they did what they could to protect themselves. Each time a letter arrived, Ovsanna recounted, "They said very loudly it was from their husbands." The girls regularly rubbed their faces with dirt in order to sully their appearance, or so the story that was handed down in their family goes.

Before long, Stepan's sister Aghavni came down with typhoid fever. She still had her baby, Louise, to breastfeed, but now she could barely care for herself. "Aghavni couldn't give milk to the baby," Ovsanna re-membered. "So they gave the baby cow's milk." Something her little body had trouble digesting.

"My aunt was so sick that her mind was troubled. All the time she was singing," Ovsanna said. The hallmark of this disease is its high fever, which usually hovers around 103 or 104 degrees. Aghavni seemed to be sliding, and she was quarantined away from her older children and relatives while her younger sister, Mari, tended to her. Delirious, Aghavni swayed back and forth, like a rocking horse, repeatedly recit-ing the song lyric "It's nice to have good friends."

Soon, the baby fell ill too.

Around midnight on that dreary night, a cattle car pulled out of the darkness and into Chai. On the platform, Armenag tried to catch his breath. With his sisters trailing him, he had sprinted there from his tent, pitched in the nearby field. He did this whenever he heard the si-ren's call, always hoping it carried his soldier brother, Stepan.

Peering inside the compartments, Armenag could tell it held a labor company, a good sign. But he couldn't find his little brother. Had the coachman Khoren given him false information? Stepan should've been calling him by now, shining a light, something. Yet the cars remained still, like a ghost train. In one car, Armenag approached a work serviceman. No, Stepan was not there. Precious time elapsed. He asked another, who directed him to a different car. Still no luck.

Armenag had one last car to check. He found someone from back home who knew Stepan. Nudged awake, Stepan learned his brother was outside, loyal Armenag whom he hadn't seen in more than half a year. Already, Stepan had dozed away precious time, and now he had less than two minutes remaining before the train left. With the guards sleeping nearby, he grabbed onto the high iron window and pulled himself up. Glancing out, he could see his brother, his face inches from his. How he'd longed for this reunion, and it was almost gone. Armenag pushed a small pouch through the bars.

It had money, Stepan could see. He couldn't accept it, and he shoved it back into his older brother's hands. The locomotive hissed; it was about to depart. One final time, Armenag forced the pouch back, besting him, like a game of hot potato. "You need it more than I do," Armenag insisted. The brothers began weeping. "Goodbye," Stepan said. "Kiss everyone for me." The eastbound train rocked away, widening the distance between the two. Armenag grew smaller by the second, like a lost kite, until he disappeared completely into the expansive night. Stepan laid awake, reflecting on the sadness of it all. *Was that our last conversation?* The moment weighed as heavy as his fatigue, to which he eventually succumbed.

The next morning, Stepan awoke to feet trampling him as the others scrambled toward the train's exit, queuing up at the latrine. The train had pulled into the town of Konya, the *vilayet's* capital and home to fifty thousand residents and a camp where many families from Adabazar now lived.

Suddenly alert, Stepan had the urge to urinate. He stepped behind the others into a line that snaked around the station. The wait gave him plenty of time to set aside the forty-*para* fee and count the remaining money in Armenag's pouch. Two hundred ghurush. That was about two weeks' wages for unskilled workers; even more, now that there was hardly any income. Two hundred ghurush that his family needed also.

Half an hour later, Stepan climbed back into his cattle car, which continued eastward on the Baghdad Railway. In the near future, a passenger would be able to ride from Berlin to the far-flung Arab city. At the moment, though, the tracks dead-ended at the foot of the Taurus Mountains — the direction Stepan was headed — its craggy peaks already visible. Railroad workers were racing to burrow a tunnel, the impasse slowing down the war effort — and the deportation.

Across a plain, Stepan's train trembled down the last of the tracks toward the Bozanti station, the end of the line. Outside, the sunlight illuminated what they would soon cross: the Taurus Mountains, rising approximately two miles into the sky. Clouds usually dipped around the peak of the most western part of the Himalayan Mountain belt. At Bozanti, the troops and supplies were unloaded and then packed onto mules or carts for transport to the other side. The tracks picked up again there, almost as if the railway had been holding its breath through the mountains. Then everything was reloaded onto the trains. All of this occurred during ambushes that came on like tropical storms, and lives hinged on the speedy reinforcement of servicemen. Ahead of Stepan and his work mates, the crossing loomed, as it did for all Armenians pushed this way. All that carved rock. Steep cliffs. Bottomless gorges. Eighteen hours of walking. Not many Armenians survived the passage.

Beside towering pinnacles, Stepan's train halted. At the station of Bozanti, the guards were changing. From here, the conscripts would march. Two cavalry guards, positioned at the fore, led the way. Clutching three days of rations and a hefty pack, Stepan strode into line, and all the men were steered onto an old caravan path hugging the mountain. It was the beginning of the grim journey to the Cilician Plain and, eventually, the ancient town of Tarsus — or so they hoped. Already, some men couldn't walk. They had rented donkeys to take them across. The cost was twenty ghurush: fifteen for the gendarmes, five for the owner.

On the trail, Stepan stepped in the wake of the caravans and work battalions that came before. Some five hundred thousand Armenians would similarly make this passage. Dr. Wilfred Post, an American physician based in nearby Konya, traveled to the region. While investigating the area around Bozanti station, he observed one particularly gruesome camp. "The valley was strewn with graves, and many of them had been torn open by dogs and the bodies eaten," he wrote. More than a thousand had perished at this place, he was told, and more would soon, given the large population still struggling across the mountains and into the Mesopotamian desert. Less than 10 percent were reaching Deir Zor, their destination.

For Stepan, an hour passed on the road, then another. Ancient fortresses barbed the terrain. A castle from the Genoese period stood crumbling. Another one from the Cilician kingdom sat atop a hill. Named Lampron, it reminded everyone that this was once officially Armenian land — at least for a brief period. Now the ruins gazed out on the slopes littered with its people. In the fall of 1915, approximately ten thousand Armenians wasted away there.

En route to Aleppo, the Ottoman army's German field marshal bore witness to the Armenians' plight. "At this point, the sad view of the Armenian deportees strikes the eye. The plan is to settle them at the southern foot of the Taurus Mountains," wrote Colmar Freiherr von der Goltz, soon the German commander in chief of Ottoman troops in Mesopotamia. "However, since it was impossible to provide sufficient humanitarian care to such a large number of people, they are living in an endless misery. We are witnessing an enormous disaster involving an ethnic group."

There was no help coming to the Armenians, partly because the Ottoman government had prohibited it. In October of 1915, the *New York Times* ran a series of stories about the deportations — "Spare Armenians, Pope Asks Sultan," "Turkey Bars Red Cross" — along with reports that it was possibly too late for anyone to help: "800,000 Armenians Counted Destroyed."

Five hours into their march up the mountain, as Stepan's company passed a thick grove of trees, footsteps pounded from behind. Stepan turned to see a Turkish captain charging toward them brandishing a thick wood baton. In a rush of confusion, Stepan froze. When the captain caught up, he started swinging his stick left, right, and center. Stepan crumpled under the blow, his whole body sharp with pain; his friends did too, their screams rising into higher altitudes. "You are not advancing fast enough!" the man howled. With a thud, the club crashed down on Stepan again.

On the steep trail, Stepan scrambled into a run beside his comrades, his heavy bags listing from side to side. Stepan thought he might die if he stopped. He'd watched the fate of those who fell, pummeled so severely they could barely move. Fifteen minutes of sprinting uphill, and Stepan couldn't take another step. Gasping for air, he huddled with his friends. Other than dying, a *baksheesh*, a bribe, seemed like the only way out. *We have ended our nightmare,* he thought. The brute receded into the woods, the fifty-eight ghurush in hand.

When the sky dimmed to black and the trail before their feet melted away, the troop bivouacked. Stepan put his head against the hard, cold ground to rest. He had to keep up his strength. The bodies of those without it pockmarked the path. Sometime during this long trek with his battalion, his good friend Tevon Harutunian couldn't take it anymore and had dropped to the ground. Stepan dug a grave for the silversmith beside the path and then had to keep walking. For years, Tevon's wife, Sirvart, did not know the circumstances surrounding his death. Saving his engraved spoons, the widow carried them postwar to her new home, Los Angeles, with her three surviving children. Sirvart and Tevon's baby girl, Lucintak, or Moonbeam, also died in some nameless place on the deportation road. Sirvart made a casket from a shoebox to bury Lusintak, a moonbeam with no light.

Back in Chai, the blush of fever faded from Aghavni Miskjian's face. *Achki luyse,* the light of joy was returning to her eyes. Her tiny daughter Louise, so sick with scarlet fever, recovered too; her heart, though, would be forever scarred. Still, all were alive, thank God — Armenag, Hripsime, Arshaluys, Aghavni, Mari, Zaruhi, the husbands, the children — but each day concluded without word of Stepan. Beside the buzz of passing trains, Aghavni launched a letter-writing campaign around a single question: *"Stepan ur e?"* Where is Stepan? She asked the

Adabazartsis in her camp and in faraway ones, anyone, for news. Stories drifted in from Mesopotamia, like charred bits of scraps, about caravans disappearing into the desert beyond the mountains. Stepan had gone that way, they knew. She begged for clues, but there were none.

Uncertainty hung over the rest of the family too. One day, as Ovsanna was standing in front of her tent, a German soldier passed by and slid his hand, like a blade, across his throat. When retelling the story decades later in her longtime home of Marseille, she retraced the line on her own neck, across the pleated folds of old age, as if she were witnessing it again. She remembered other seminal moments too, like the day when word circulated about an impending visitor — the minister of the interior himself. Just hearing about this sent the camp into a frenzy. "Talaat [Pasha] was going to come," Ovsanna recalled. "Everyone scattered around, some left, some right."

With Chai and neighboring stations littered with the infirm, military authorities worried their diseases would spread to passing troops, and they asked to close the camp. In an encrypted telegram, the province's governor promised that the two thousand Armenians at Chai would soon be moved on.

The Miskjians didn't want to go. Ovsanna's father bribed the guards to let them stay, but on a subsequent evacuation, even *baksheesh* had lost its power. The extended family rolled up the beds, tore down their splintered homes, packed up the pieces. To Ovsanna, it felt like they'd spent two years there, which it might have been; the exact date of departure is unknown. Without choice, most of them hoisted up their belongings and started southward to the nearby town of Bolvadin. Ovsanna and her parents and brother, however, found themselves forced in a different direction, and the Miskjians were split up once more.

The Interior

2007

"WE NEED TO LEAVE," I whispered to Baykar. Down an alley, police officers in an SUV stared at us.

An hour earlier, we had arrived in Bolvadin, set at the base of brown mountains deep in the country's interior. As we wandered the narrow streets, passing restaurants, mosques, and newsstands, I tried to get a feel for the place by snapping some photos. But after a few clicks of my camera, a veil of silence had fallen around us. The outdoor café, which had been joyous with banter a moment earlier, had hushed, the men's heads turned in our direction. Just a few women paced the sidewalks, I belatedly realized, and all were covered up, unlike me, with my exposed arms and my hair long and loose.

I had heard the region was conservative, and now I could see how true it was. With each step we took, the hymns of the prayers seemed to follow us, amplified through the hot air like music. Very different from Eskishehir, a modern city south of Adapazarï where at night twenty-somethings packed the bars by the lit-up river, downing beers and *meze*. East of there, the landscape seemed to crack from thirst. Swaths of nothing had rolled by, then a burst of green vegetation, then nothing again, alongside mountains. I remembered my mother always talking about this region, churning the syllables like wind-whipped tumbleweed. "The Turks took all the Armenians into the *interior,*" she would say. "They took both Mama and Baba."

Bolvadin was larger than I'd anticipated but surprisingly similar to how my mother's cousin, Ovsanna, remembered it from the deportation's end: full of mosques, and no gardens. My grandfather's family was lucky enough to stay in a shed here following the closure of the camp in nearby Chai. Earlier that morning, we'd disembarked in Chai and observed the other passengers arrive to the embraces of loved ones. I thought of my grandfather's own fleeting reunion there with his brother, under very different circumstances. Beside the tracks stretched a field, green with grass. The former site of the camp, I surmised, or nearby. Their prison without bars.

Now, in Bolvadin, Baykar spotted the policemen in their uniforms, but he didn't find their glare unnerving or worrisome. Still, I did, and ushered him away from the square. "Let's find a cab," I said, and we flagged one down.

"Afyon," I said to our driver — back to our hotel. I was tired from scouring these towns like an amateur detective, on the hunt for traces of my family's time here. Sinking into the seat, away from the disapproving eyes of the men and the policemen in the square, I felt myself relax. I pictured the next day and our visit to the famed mineral springs for a short respite. After a handful of miles, Baykar and I started chatting with our driver, trading names, marital statuses, and ages. The driver revealed his to be sixty-five.

"You look forty-five!" I exclaimed. "What is your secret?"

"I don't drink alcohol and am very religious," he replied. In the 1940s, he was a football player, Baykar translated, then he had turned to honey production.

"I'm going to eat a lot of honey now that I see how young you look!" I said. We all laughed.

"Come to my relative's wedding as my guest," he said. "All of us Christians and Muslims are brothers. I am your brother." I smiled as fields, small factories, and the mountains passed outside the windows.

"Why have you come here?" he asked. He didn't seem suspicious, just curious. Understandably so, as this was not a typical destination for American tourists, unlike the bikinied Aegean and Mediterranean beaches. We were in Turkey's dead center. I hesitated to answer his question, and then I decided to tell him. I had to, if I wanted to understand the current Turkish sentiment. I started with my ethnicity.

"*Ermeni?*" he repeated, just to be sure. I nodded. "Good," he replied with a broad smile. "I am your uncle."

Likening me to family seemed like a good thing, whatever the relation. *He's letting me know that he still welcomes me.* Just to be safe, though, I also told him about my father's ancestry, thinking it'd be more palatable somehow. Then he surprised me by passionately expounding on the unsuccessful British attack on the Dardanelles in 1915, one of Turks' greatest military victories. "There is no problem," he concluded. "Each year, in the Dardanelles, there is a British ceremony."

Stupidly, I hadn't thought that my Scottish ancestry might also be a minefield. Intrigued by his perspective, though, I returned to the subject of the Armenians as delicately as possible. "Why do you think some Turks don't like Armenians?" I asked. "And why do some Armenians not like the Turkish people?"

"No problem," he said. "That is finished. These happenings are old. I am your uncle." His lips no longer held their upright crescent.

How different we felt. It was over for him. For me, it was just beginning.

"During the Ottoman times, the Armenians and Turks lived together peacefully, like brothers and sisters," he continued. "When the Italians, British, and French occupied [the Ottoman lands], the Armenians had been encouraged and attacked the Turks. The Turks in return responded. There was no such thing as genocide!"

He paused. "We are good people, the Turks," he said. "We are not bad. The Europeans don't like the Turks. They say that Turks are bad."

The cabby was referring to the fraught debate over Turkey joining the European Union. Some countries were trying to make membership contingent on the Turks recognizing the genocide. The nation's poor record on civil and political rights was also hindering its inclusion, as was its persecution of writers, like Nobel Prize winner Orhan Pamuk, who had spoken about the Armenian killings.

With the rise of nationalism, I'd been ruminating about what it meant to be Turkish. Were you Turkish if you were born in that country? Did that mean my mother was Turkish? Or did being Turkish have to do solely with one's lineage? A few hours earlier, Baykar and I had been at a restaurant in Chai, marveling at the quaintness of the Alps-like village. The establishment's owner kindly expounded on all the local attractions and gestured to the mountains, dense with trees. When Baykar remarked that he was from the capital, the owner's tone had shifted. "I couldn't live in Istanbul," he said matter-of-factly, "because there are too many Kurds." The Kurds are the new Armenians, I realized; the discrimination against this minority and their struggle were widely noted. But their role in the persecution of others, namely the Armenians during the genocide, was less well known. Despite the Kurds and the Turks' shared faith and citizenship, I didn't know if the Kurds would ever be Turkish enough. After all these years, Turks still seemed to prefer a country of only Turks.

Ahead, a massive outcropping of rock appeared, rising suddenly like an island out of the sea. We were nearing Afyon. The city's Ottoman name, Afyon Karahisar, meant "Opium Black Castle," which sounded more like a drug den in a Bond film than the decaying structure I could see sitting atop its pedestal. After looping around, our driver deposited us at the train station. As he sped away, I noticed the same SUV from Bolvadin parked across the traffic circle, a policeman in it holding a camera. *Oh no,* I thought, *it's them.* I stiffened with fear, my palms sweaty. "Don't be obvious," I said to Baykar, "but officers are taking a photograph of us. They look like the same ones from Bolvadin. I think we're being followed." Again, Baykar seemed unconcerned, used to the ways of the police here.

That evening, while Baykar and I were eating dinner in Afyon, a hotel photographer approached our table and asked to take a commemorative picture. I shook my head back and forth to say no. A few minutes later, he snapped a picture of a child nearby, with us in the background. I turned my head so as not to be in the shot. Was I being paranoid? Had I watched too many spy thrillers? All I knew was that I was scared, that I didn't want to be questioned, and that I had to leave. We'd have no day of rest.

Later that night, I pulled up the covers on my bed and buried myself underneath, as if this would protect me from surveillance, from my own delusions. We had to change our itinerary, I decided. We'd go by bus, not train, to our next destination, Konya. From there we'd take the rail again to Adana, a city southeast of the Taurus Mountains — or Toros Daghlari. That was where my grandfather had been forced to climb, while being beaten with a stick. Scared, I remembered another maxim he wrote, and it calmed my breath: "Man should always be ready to confront every day and at any hour new hardships and misfortunes because these form an inseparable part of life."

Infidel Mountains

1915

AT LAST, Stepan could make out the faint outline of a town: Tarsus, on the mountain's other side. The crossing was behind him now; he could breathe easier. Ahead, the thick vegetation parted in a clearing. A few more strides and he could see one thousand Armenian laborers in a field near the range's base. A chain of guards now enveloped his battalion, scrutinizing their every move. There had been more Armenians, he'd heard, dispatched elsewhere to "construct roads," or so the officials had said. Listless, he sat there all afternoon, as the shadows grew longer and longer around the trees. *Our situation is deteriorating,* he told himself. *I have to flee. I have to find a way.*

It was sometime in October now, or possibly even November. In the ensuing darkness, a disembodied voice made an announcement: The guards needed five *nefers,* or privates, to fetch some bread. Stepan volunteered; this was his chance. With his experience as a major's servant and his gift as a salesman, he persuaded them one of the privates should be him. At first light, he straddled a horse and departed alongside the four others and the pack animals, all lumbering down the slope into a marred landscape. With a warmer climate than the mountains, orchards and gardens normally graced Tarsus, part of Adana *vilayet.* There were some cotton fields too. In wartime, though, the plants

coiled brown and shriveled. Down the streets, the horses clopped past
the train station, where the rail had picked up again. It was packed.
Hordes of Armenians shuffled about; some twenty thousand of them
languished in the area that fall. Stepan scanned the crowd. *Wait, who is
that? And that?* Some friends from home, Adabazartsis. He dismounted
and rushed to them. In telling him about their harrowing experiences,
they confirmed his deepest fear: *Armenians are being left to rot.* He couldn't
ignore the mounting evidence, and he ticked off the facts in his head:
*Our bread ration is distributed very irregularly; they don't take care of the sick; the weak
are discarded on the road.* He'd join these friends, desert his company. But
first, he needed bread. Even a healthy young man couldn't make it far
without that.

Throughout the region, the misery wasn't abating. Edward Nathan, the
U.S. consul of the nearby town of Mersina, watched the tragedy unfold.
"The stream of deported Armenians from Anatolia to Syria continues,"
he wrote to the U.S. State Department. "In enumerating the various
distressing elements connected with this movement, I perhaps failed
to point out the terribly unsanitary conditions that prevail in the vicin-
ity of the camps or stations near Tarsus and Osmania. These result in
part from their overcrowded state, but largely also from the imperfect
burial of the corpses of the victims of starvation and disease. The mor-
tality among the deported is daily increasing in percentage, and, when
the rains set in, the toll will be frightful. The feeding problem is completely neglected, and will become worse in the future."

The season was turning colder, an unfortunate development for the
Armenians and the combatant troops. Continental Europe was alight
with firefights, and the campaign in Mesopotamia was continuing to
intensify, with the British advancing toward Baghdad.

In the Tarsus bakery, Stepan placed his battalion's large order. The
bread wouldn't be ready until midnight, hours away. Only after that
could he flee. Stepan was curious about the town, important in biblical times and the Middle Ages. He remembered learning about it from
his history teacher in elementary school, and he began to stroll through

it with the other recruits. For a Christian like him, the place had a special meaning; it was the birthplace of Paul the Apostle, one of Jesus Christ's most important missionaries. Meandering, he would have seen St. Paul's Church and perhaps even passed under the arched monument, St. Paul's Gateway. The town was also known for its amorous past. In 41 B.C., Cleopatra visited to tryst with Mark Antony, though there was nothing romantic about the place now. The covered bazaar reminded Stepan of the one back in Constantinople, tentacled across narrow, dirty streets.

The light continued to fade from the sky, and eventually blackness replaced it. At midnight, he filled his sacks with the warm bread. After climbing atop his horse, he rode back to the field. As his horse trotted, his thoughts turned toward escape again. Mixing into the teeming deported villagers, flowing around railway terminals, seemed an uncertain future, but at least it was a future, however short. First, though, he had to fetch his bag and his allotment of bread. Back at camp he retrieved both, and then he scanned the night for opportunity. The previous evening, he had noticed a few guards who were careless in their positions. He rose and crept toward them. He crept some more, then froze. More sentries were pacing the perimeter, and they brandished their lethal instruments at their sides. Not a sliver of an opening — or any hope that he'd get out.

He retreated and ate his ration in silence, in hopelessness. In the morning, new marching orders sent the company to Tarsus. Somehow, Stepan and the one thousand soldiers all fit on one train. Stepan was shoehorned into one car amid sixty others, their bodies crumpled like pages of a newspaper. They had no idea where they were headed. Stepan's train shook down the track, farther into ancient Cilicia.

Past the city of Adana, the train chugged onto a plain bookended with rises. Soon, the shifting scenery unveiled a river and the old fortress of Toprakkale. Stepan could feel the train slow. Outside stood the station of Osmaniye, home to another camp, where one witness had spotted fifty thousand deportees spreading out a half a mile in both directions. The gendarmes were merciless custodians, repeatedly hitting the hungry and thirsty refugees. That fall in Osmaniye, only about one-third had enough food and countless numbers suffered from typhus.

The farther east Stepan journeyed, the deeper he seemed to sink into a godless land. The camps became bleaker, the wretchedness unyielding. For the next thirty minutes of the ride, the tents and homeless continuously shadowed the rail.

A Swiss nun rode her horse through this corridor in November 1915 and documented what she saw. "Thousands of exiles are lying out in the fields and streets without any shelter and exposed to the power of any brigands," Sister Paula Schäfer wrote to Henry Morgenthau.

> Last night, about 12 o'clock, a little camp was suddenly attacked. There were about 50 to 60 persons in it. I found men and women badly wounded, their bodies cut open, with broken skulls, or in a terrible condition through stabs with the knife. Fortunately I was provided with clothes, so I could change their bloody things and then bring them to the next inn, where they were nursed. Many of them were so much exhausted from the enormous loss of blood that they died in the meantime, I suppose. In another camp we found thirty or forty thousand Armenians. I could distribute bread among them. Desperate and half-starved, they fell upon it; several times I was almost pulled down from my horse.

The sister was fortunate to have been allowed to hand out any bread at all. When Morgenthau asked permission to allocate one hundred thousand dollars' worth of aid, he was refused. "It is the opinion of Enver Pasha that no foreigners should help the Armenians," explained Halil Bey, the minister of foreign affairs. Indeed, on November 1, 1915, the *New York Times* seemed to indicate the policy was official: "Aid for Armenians Blocked by Turkey: Attempts to Send Food to Refugees Frustrated, Says the American Committee."

Unaware of the international attention, Stepan was focusing on his immediate surroundings. The Amanus Mountains were sharpening into focus, going from a fuzzy image to a range fluted with highs and lows. At the foothills, his train halted. Outside stood Mamure, the last station. A command post for the Baghdad Railway workers, it was also home to another encampment of Armenians, with their pleas to pass-ersby: *"Hanum* [bread], *hanum,* I am hungry," and "We did not eat any-thing today and yesterday!"

Stepan and his troop filed off and learned there'd be no rest. Uphill again, they marched a footpath twisting up the slope. In two hours, they reached the top, where there was a glade stacked with piled wood. In another moment, more logs came crashing down, as the Armenian la-borers already there sawed down trees with antiquated tools. With the coal shortage, the trains needed the wood from this area to run. Com-pleting the tunnel was difficult, especially since the mountain lacked a natural passage, like the Taurus's Cilician Gates. Blasting a hole through the Amanus was becoming one of the railroad's costliest efforts. An opening in September had allowed for minimal service, but that was all. High on the mountain above, a foreman now explained that Stepan and the others had to erect massive posts, around thirty feet long. Each one weighed two hundred and eighty pounds, too heavy for a man to carry alone. So the thin *emanetji* reached down and grabbed an end of one while another serviceman did the same, and the two heaved it down the steep grade. *Survival of the fittest,* Stepan told himself, and he was thankful this assignment didn't last long.

With four days' rations of bread, the men were now led to a tapered trail. It was time to cross the range. At the base of the ascent dipped Kanlï-Gechid, a gorge and waystation for the terrified Armenians who had camped there. All night, the animals screamed and howled.

On the path ahead, something blocked Stepan's way. It was large, like a deer, but motionless. Closer inspection revealed an old man, his eyes staring out, dead as the day. Aghast, Stepan and his work mates gaped in disbelief. In a few more steps, they came upon the body of a young mother, and they trembled in horror. Next were some children, their small hands and feet frozen in rigor mortis. These were the first civilian deaths they'd seen, but they would not be the last. As the men climbed farther still, and over more corpses, the faces of death began to look the same. They were the remains of the hundreds of thousands preceding his convoy. Strange what the mind could become accustomed to, Stepan always marveled. All along the grueling march, more of Stepan's fellow workers succumbed, the guards barely taking notice. *Nobody cares for us,* Stepan thought. *Only that we bury their men and continue to advance.*

Dutifully, he trudged forward anyway. He had to keep his wits about him, concentrate on when to expend his energy, when to conserve it. The area was nicknamed Giaour Daghi, the Infidel Mountains, for the large Christian population that had resided there over the years. Up ahead on the hillside stood an Armenian church. *Like a fortress,* Stepan thought. He was approaching Hasanbeyli, a picturesque hamlet some seven hours by foot from the bloody gorge. He remembered reading a story in the press debating its future: Preservation or demolition? The latter had won out. Not long ago, Armenians used to agonize over issues like this. He slept beside a brook that night and trod through the village the next morning, his sorrow growing as he walked. He couldn't find any Armenians there, not a single one. Armenian script was still engraved on the storefronts, the ghosts of its recent past. Turkish shopkeepers, unfamiliar with the Armenian merchandise, now sold items at an inflated cost.

On the track ahead of Stepan, some Arabs approached, dressed in the real uniform of Ottoman soldiers, not in civilian rags like him. Fresh recruits, he surmised, accompanied by five gendarmes. As the two platoons crossed, the recruits lunged forward and attacked the depleted Armenians with a fierce force and strength. Stole nearly every *para* from Stepan's work mates that they could find, the entire assault under the guards' watch. Thankfully, they hadn't discovered all of Stepan's cash. Still, he wondered, *Had they ordered this?* When there was nothing left for the soldiers to take, Stepan and his platoon struggled onward, eventually cresting the mountain. If they weren't so beleaguered, they might have felt like kings. A magnificent panorama stretched before their eyes; the view of another expansive plain. With each stride, they descended Infidel Mountains, eventually reaching Baghche, the terminus of the railroad's new tiny tunnel. Stepan was entering Arab country, stepping closer to the the Syrian desert — and Deir Zor. His heart weighed heavier as he walked, as he realized the truth.

The situation has completely changed, he thought. *They're not taking us away as soldiers; they're taking us away to annihilate us.*

PART THREE

Red River

Red River

The Headscarf

2007

THE SALESWOMAN BECKONED ME TO SIT ON A STOOL next
to a glass case. Two other ladies circled me and began to smooth my
frizzy hair into a bun. In this cramped clothing store in eastern Anatolia, I was having a spontaneous makeover in order to visit one of the
city's oldest shrines, the Yagh Jami, the Oil Mosque. The evening before, Baykar and I had arrived in the large city of Adana after crossing
the Taurus Mountains by rail. The chiseled rock and cliffs reminded me
of Yosemite, a wonderland whose heights my grandfather had struggled
to cross with his labor battalion.

At the store, the nice woman took out a white cotton bonnet and
pulled it down snugly on my head. In the back, she crisscrossed the ties
into a bow. Briefly, I felt like a polygamist sister-wife from Utah, but
she was far from finished. She folded the orange-and-white scarf I had
picked out and twisted it. She needed a pin but didn't have one, so she
reached into her head covering, removed her own, and tenderly tucked
it under my chin. I was deeply touched by this small act of kindness, and
I smiled wide. She responded in kind.

Ever since leaving Istanbul, I had worried about what each day held,
even more so since my interlude with the Bolvadin police. I felt like
I was doing something wrong, even though I wasn't. Now, between
these racks of colorful clothes, I felt coddled, cared for, and safe. As the

Turkish woman gathered the scarf tightly at the nape of my neck, one of the other saleswomen seemed increasingly bewildered by my visit. Since Baykar had been patiently translating our exchange, she turned to him.

"Did she become a Muslim, or she will? You teach her the Kelime-i Shehadet." She was referring to the Muslim declaration of faith.

"I don't know it either because I'm Armenian," he replied.

"Really? Truly?"

"Yes."

"Where do you live?"

"I'm from Istanbul. My friend is from the United States."

"Why does she want to wear a scarf?"

"To enter the mosque."

"The foreigners like your friend show more respect. I saw Turkish girls in mini-shorts and no head covers."

Neatly, the other woman arranged the pleats to fall like a fan across my neck. She gestured for me to look in the mirror. I stood up and glanced at the image staring back at me. With my olive skin and dark eyes, I resembled the ladies in the store, only slightly younger, with chipmunk cheeks. The scarf matched my cream long-sleeved shirt and chestnut ankle-length skirt. "You look good," said the saleswoman. "Let this bonnet be my gift to you." With the circle of women admiring my new style, their eyes full of pride, I felt like I was about to leave for prom; and my date would appear any minute.

"Perform also the *namaz*," said the third woman, "and become a Muslim and profess your faith by pronouncing the Kelime-i Sheha-det: *La ilaha illa-Allah Muhammadun rasulu-Allah.*" That was a holy prayer, I learned later; it means, "There is no deity but Allah and Muhammad is the messenger of Allah." Touched by their generosity, I thanked them and exited. Finally feeling presentable, I was ready, and I crossed the courtyard to enter the red-roofed, five-hundred-year-old mosque.

At the riverfront an hour later, Baykar and I watched the water lap under the arched Roman Bridge; a creation from the time of the Roman emperor Justinian, it resembled one I'd visited outside Adapaz-arï. That mid-August day was hot, even more so because I was wearing my hijab. The cool of the shady trees in the slender park offered a temporary reprieve. Near us on a bench sat two American soldiers in uniform. With all my uncertainty about being in Turkey, I was happy to see them, though given my transformation, they didn't notice that I was a foreigner like them. I wasn't surprised by their presence. Eight miles away was Incirlik Air Base, where the United States launched much of its operations to the Middle East, with the cooperation of its Turkish ally. Before my trip, the media had frequently mentioned that U.S. politicians feared the Turks might cut off access to the base if the Americans recognized the Armenian killings as genocide.

It topped the news because this year there was momentum for official recognition in the U.S. Congress with Democratic control of both houses. The new Speaker, Representative Nancy Pelosi, had promised to move a vote on the genocide issue to the floor. My mother was ecstatic, the Armenian community aflutter, but I remained doubtful. Already, the administration of President George W. Bush was vigorously opposing it, given the wars in Iraq and Afghanistan and the strategic importance of Turkey in the region. And soon eight former secretaries of state weighed in too, cautioning that passage of such a bill would "endanger our national security interests." Sure enough, when a House committee approved the resolution two months later, Turkey recalled its ambassador from Washington and hinted at further action. "Unfortunately, some politicians in the United States have once again sacrificed important matters to petty domestic politics despite all calls to common sense," Turkish President Abdullah Gül said following the vote. Not surprisingly, the resolution died. Coincidentally, the base had been built partly on land once owned by Armenians — until the genocide.

After two days of exploring the area, Baykar and I left Adana. We'd moved up in the world and were traveling by car now. Behind the wheel sat an avuncular Kurdish man named Jemal who would drive us the last leg to the Syrian border. Topping six feet, he was a hulking figure, but with his sunny countenance, he was all beard and warmth. In the back seat, as the green countryside passed outside my window, I clutched my bonnet from the store and thought of the friendship that could still grow between Armenians and Turks if given the chance. Jemal was now about to take us up the winding roads of a range once called the Infidel Mountains. Looking out, I couldn't believe how here, in this peaceful forest, my grandfather had stepped over corpses; the tranquility and the death seemingly incongruous. Jemal pointed to an old caravan trail off to the side of the road. I asked him to stop, and then I began to walk where my grandfather had nearly one hundred years earlier. The dirt was soft and powdery, the air thin as I climbed. In minutes, my lungs burned and my calf muscles ached, just like in Istanbul. And I was hiking with food in my stomach, without my life's possessions on my back. After an hour of huffing and puffing—and cursing myself for my lack of fitness—I gave up and hopped back into the car, a choice my grandfather didn't have.

Farther up the road, we reached the quaint village of Hasanbeyli. Jemal directed his big yellow van down the main street and then up a hill, stopping in front of an old stone blockhouse. Raising his arm, he pointed. I looked but didn't see anything. He told me to try again. Finally, I saw faint writing—in classic Armenian characters—tucked under the eaves. *Blessings and abundance will be in his house*, it read, from Psalm 112. I wondered what had happened to the owners, remembering the sorrow my grandfather felt at stumbling past the suddenly emptied town, the goods of the Armenians sitting on the shelves. What had happened to their abundance, their blessings? Jemal knew this stop well. For years, he had been guiding Armenian tourists through eastern Turkey to visit the areas their families had fled from or where they'd died, to view the homes that sometimes still stood, to see places like this, lost in time.

Earliest surviving photograph of Stepan Miskjian (pictured left) and his friends, taken circa 1910 in Adabazar, then part of the Ottoman Empire.

When in Constantinople for business, Stepan stayed in a boarding house on Mahmud Pasha Street, in the old Stamboul district.

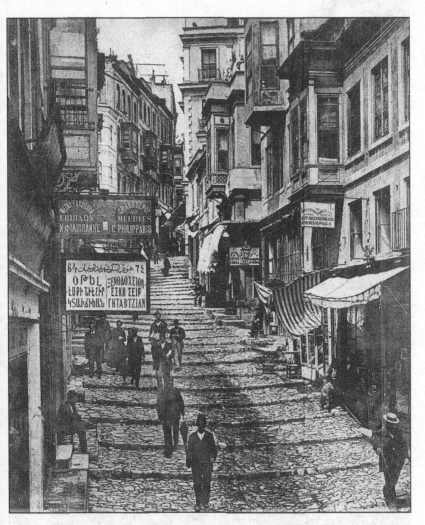

Stepan would regularly carry his heavy deliveries up this steep street in Constantinople, past Armenian, Greek, and Turkish shops. In 2007, Dawn huffed up the same street and discovered that, not surprisingly, she had less stamina than her grandfather.

rue de tchark Ada-Bazar

الطه بازارلی چرج جادهسی

A typical street in Adabazar at the turn of the twentieth century, with homes similar to the Miskjian family's.

Gare de chemin de fer, Ada-Bazar

الطه بازارلی استاسیونی

The train station in Adabazar. Stepan was arrested by the police here in 1913 and wrongfully accused of transporting illegal weapons.

Soon after the roundup of intellectuals in Constantinople in April 1915, the mass deportations began. Here, armed guards lead Armenians out of a town in eastern Anatolia that spring.

Like Stepan's family, these Armenians were forced from their homes and then deported across Anatolia by train. "They crammed 880 people into ten cars," Franz Günther, an executive with the Baghdad Railway, wrote to the company's chairman on October 30, 1915.

The hard labor of road construction often fell to Armenians, including Stepan. Here, Armenian men toil in the Taurus Mountains, which Stepan crossed on foot in the fall of 1915.

Women and children made up the bulk of the Armenians in these desert camps. To withstand the elements, many layered their clothing, and much of it was worn through shortly after the deportation began.

Marched to the Syrian desert, an Armenian mother mourns her dead child.

Human bones found by farmers outside Ras al-Ayn, another slaughterhouse north of Deir Zor, Syria.

Dreams Traded for Bread

1915

THE SILHOUETTE OF A MOUNTAIN stood out in the waning light. Stepan was marching toward it with the remaining Armenian soldiers. Up ahead, he could see a mass of people, perhaps tens of thousands, camped beside the train station under a rainbow of blankets and tents. *More Armenians,* he thought. It was cold, perhaps November, when Stepan and his labor battalion came on Qatma and the train station of the same name.

He was only thirty miles from the large city of Aleppo where American diplomats and German teachers strove to save the Armenians, but he might as well have been a continent away. As the conscripts drew near, thousands of deportees began to rush toward them, searching the faces of the six hundred men, hoping to find a lost brother, father, or child. Many watched the tide of refugees continually wash in each day, carrying more uprooted Armenians, or at least the ones not discarded on roadsides like tangled kelp. Stepan was thinking about his own losses. Since Tarsus, almost half of his comrades had been left behind, most of them dead.

He scanned the crowd. The ragged Armenians were dressed in civilian clothes, same as him. He wore nothing to distinguish him, no uniform or insignia on his sleeve. Excitement kindled in him. Among them, he wouldn't be spotted. He could disappear. He examined them closely; perhaps, he'd find someone he knew. Just then he saw, amid all deportees, hirsute with overgrown hair, an Adabazartsi. He continued walking, and he spotted another, this one a neighbor.

Thrilled with the secret of recognition, he continued in step with the others like a dutiful soldier, all the while plotting his escape. As Stepan's company traveled the last bit, the guards divided them into two neat lines. Mounted, the sentries galloped back and forth, the hooves of the horses slapping the ground, creating a stuttered rhythm to the scene. *Clip-clop. Clip-clop.* It was evening. The camp's Armenians had pressed closer to Stepan. *Clip-clop. Clip-clop.* A woman he knew from Adabazar made eye contact. *Clip-clop. Clip-clop.*

Now. Swiftly, he took the bag off his shoulder, handed it to her, and, in the same sweep of motion, left his column. Pulse pounding, he melted into the crush of deportees, the glare of the gendarmes mercifully elsewhere.

Setting out toward the sea of tents, sprawled some three hundred yards away, he marveled at the ease of his second escape. He was no longer a soldier with rations but a deportee, flung to the empire's fringes. It took a moment to absorb his new identity, cemented when his battalion marched out a few hours later without him. Escorted by armed guards, the regiment was veering southeast, toward Aleppo, where he could only guess their fate.

As evening fell, Stepan explored the camp. Beside the train station was some type of military depot. In the adjoining field, small glows of fires lit up the maze of tents, the smoke rising. No trees in this forsaken spot, only patches of thistle. He ran into the slipper maker Mardiros Kesoghluyan, his comrade from the *amele taburu,* his labor battalion, and brother-in-law to the late Tevon. Mardiros too had escaped, but to where? Not to freedom, that seemed certain. Multitudes of guards patrolled the circumference of the camp. The destitute were strewn upon the ground like detritus after a fair. Stepan walked past them, their once-plump bodies starved down to bone. What kind of place had he entered? He didn't know. His dire new circumstances settled on him like the darkening night.

At daybreak, Stepan waded deeper into the muck. Whole rivers ran through the field, flowing with raw sewage and trash. He stood on its bank, the air, at the time, thick with stinging flies. All around, people retched, their bodies covered in welts. No latrine anywhere, the stench overpowering. Stepan's mind churned as the difficulty of surviving this hit him. Winter was blowing closer, and his resources, including his brother's gift of cash, were already diminished.

He fingered his few coins, ever thankful thieves hadn't discovered them. Soon he'd have to figure out how to make more. He purchased some materials for a tent and began searching for more people he knew. The ones from his hometown would be together, given how they were all informally grouping themselves that way. With some wandering, he found the new Adabazar, distinguishable by its fringe of rag carpet, a thick weave specific to his region — the thread that connected them all to home. He spoke to the residents now; almost all had been robbed. He looked at them. Their features were familiar but distorted by privation, the way a funhouse mirror reflects a face. In a small clearing, he dug poles into the ground, draped a covering on top. Next, he gathered some clothes and arranged them in the narrow shape of his body. Now he had a bed. The *emanetji* lay down on the bunched-up garments, shut his eyes, and tried to push the fear away. Compared with others, he could count his blessings. Somehow he still had his pants, his shirt, his health, and some ghurush.

Not far from camp, the British and the Indians had pushed to within twenty miles of Ottoman Baghdad but had exhausted their supplies. Facing a reinforced Turkish army, they retreated back to Kut al-Amara, the latest combat zone on the Tigris River.

At the end of that week, Stepan followed the length of the stark hills in search of work. He'd heard there was water in this direction. Someone had told him the camp's lone spring had a name: Dermin's. Now, after a forty-five-minute hike, he could see it hiccupping and gliding over large stones, clean, clear, and refreshing. Five gallons weighed more than forty pounds, and some Armenians were capitalizing on the hardship. He could see them, stooping over the creek, filling jugs, and loading them onto donkeys to sell. He would join their ranks.

Scouring the ground, he found four abandoned tin gas cans. He scrubbed out the slimy residue and spent 170 ghurush — about half his remaining money — on a donkey. As he led the beast through camp, the thirsty and weak flocked to him. He sold his first can for thirty *paras,* the next for ten *paras* more, his take-home pay growing to ten ghurush a day. Not everyone could stand the rigor of the work. One friend, a photographer from home, tried to accompany him but soon quit, barely able to carry the sloshing canisters. Accustomed to such loads from his courier days, Stepan pressed on and expanded his customer base, approaching Antranig Efendi Merjanian, the respected Adabazar moneylender. Stepan thought of the last time he'd solicited him, back when he had had the promise of a future before him. Now he sold water to the dying. Similarly separated from his family, Antranig Efendi had become convinced of a terrible truth — that the Turks were trying to wipe out the entire Armenian population. The idea was beyond belief, but there was no other explanation for the bodies stiffening around them. Stepan had come to the same grim conclusion.

Still, Antranig believed his sons, army officers, would rescue him. When each new day came, and he remained in the camp, he held out hope for the next day. The parade of caravans continued to drag villagers into Qatma from Anatolia and Cilicia, the news already confirming the worst: massacred Armenian soldiers lay in ditches all over.

During the war, the empire had become overextended; it was trying to simultaneously vanquish Egypt, protect its borders, beat the Allies in Europe, and oversee the Armenian deportation. Beside the tents, the military had established Qatma as a communication center. At the edge of Arabia, it was the perfect location for logistics, sitting at a crossroads of the Baghdad Railway and a major Aleppo thoroughfare. Lines of Armenian battalions and platoons kept marching through Qatma, like ants, endless and in formation. On one particular afternoon, a labor battalion of Armenians entered the camp, trudging in just like Stepan's had, close to the refugees.

"Are there people from Adabazar?" asked one of the battalion's soldiers.

"*Emanetji* Stepan," someone replied.

Baron Khoren Mkhjian of Adabazar overheard the exchange. He hadn't seen Stepan since they had boarded separate battalion trains near Constantinople, when Stepan had asked Khoren to inform his family in Chai that he'd soon follow. Scanning the crowd now, Khoren passed his bag to an Adabazartsi lady who stood among them and then dashed to find the *emanetji*. Incredibly, he found him. The men embraced, kissing each other's cheeks. "I had no intention of fleeing," Khoren said, "but as soon as I heard your name . . ."

Stepan led him to his tent, and the two men squeezed inside. With golden stalks of straw, Stepan arranged another bed and then prepared some food. Over a meager meal, Brother Khoren decided to join him in his trade, and the two set out to the stream the very next morning. With Khoren, Stepan felt like he had family again. With Khoren, he wasn't alone anymore. One couldn't really survive out here alone.

Each day, the chance of seeing another long-lost friend increased as more refugees stumbled into the meadow, already teeming with people and tents, the population swelling at one point to more than one hundred and fifty thousand. To reduce the size, the mounted guards would initiate a *sevkiyat,* or dispatch. Stepan could now recognize the signs; the sentries burst into the tented quarters without warning, barbed with their typical shouts: "*Kalkin! Chikin! Yikin!*" Get up! Get out! Tear it down! The ones who had been there the longest had to go first and make room for the new caravans that never seemed to end. Sometimes the refugees were herded southward toward Aleppo, other times eastward along the Euphrates River, the banks now dotted with Armenian encampments.

Stepan and Khoren's water-carrying days were long and grueling. While making rounds one afternoon, Stepan felt a sudden pain in his abdomen. He raced to the nearest patch of dirt and relieved his bowels of the mess, but it soon happened again. He tried to stop this unstoppable force of diarrhea but lost the battle a few times a day, then eight, then ten. Now carrying the canister was impossible. Folk medicine in Adabazar dictated that he sit on a hot copper pot to cure this type of ailment. Without one, he turned to other homeopathic remedies, to no avail.

Stepan never knew exactly what he had, but it seemed like his neighbors suffered from the same illness. He worried the spring was contaminated, that it had polluted the water he sold. He had good reason for concern; the animal carcasses and excrement were piled everywhere, and the deportees drank from and bathed in the same source, prime conditions for the bacterial infection cholera. With all the typhus and dysentery, the camp had become an outdoor ward, though one with neither doctor nor medicine. At first, fifty to sixty new bodies piled up each day, and that number soon rose to two hundred. So many dead that Armenian men heaped them in oxcarts and wheeled them away to mass graves. As one deportee, Elmasd Santoorian, described the camp, "Qatma may have looked like Hell itself, but it actually was only its antechamber."

For Stepan, each day dimmed darker than the one before. After Khoren cooked some *chorba*, soup, and left for work, Stepan struggled to swallow a few sips of the rice and lemony broth. *My medicine*, he thought, determined to recover. First unable to walk, he now couldn't sit up on his own. He drifted in and out of sleep to the wails of the other sick Armenians camped nearby. He couldn't fight the disease, or move. The twenty-nine-year-old was sure of his future now, his days consumed with one thought: *I'm going to die.*

The man seemed to be looking for someone. He had already combed through the big tents, nearly missing this tiny one pitched at the very edge. He wondered if anyone could live there, and he bent at the opening and peered inside. A cadaverous figure lay there. The man poked his head in to get a better look at the high cheekbones and square jaw carved by illness. Something about this face looked familiar. So he spoke a few words. The figure stirred and opened his eyes. "Partik Akhbar," Stepan said.

"God, Stepan, is that you?"

"Yes."

"What's happened to you?"

Partik Akhbar Unanian entered the tent. He crouched beside Stepan, his good neighbor from Adabazar.

"I was a soldier," Stepan murmured.

It was as logical a place to start as any. So he began there, with the war's outbreak and all those hours grinding away on the road. How he'd been separated from his family. Now they were in Chai. Or were they? Months had passed without any news, and he dreaded the worst.

"I'm ill and I'm not getting well," Stepan pronounced with finality.

His neighbor's eyes drowned in tears. Stepan understood why. He didn't resemble the person Partik Akhbar had known since he was a child. The friend to his own son, the bullish *emanetji* who slogged great distances with satchels of heavy goods, always roaming the streets in his off-hours with his entourage, en route to another practical joke. Headstrong, but a boy at heart. Now he was a shadow of that man.

"Let me take you to our tent," Partik Akhbar suggested.

"Voch," Stepan replied. No.

Partik Akhbar insisted again, but it was no use; Stepan would not budge, not wanting to be a burden to the man's family. Disappointed, Partik Akhbar left. Not long after, two girls appeared at Stepan's tent. He recognized them as the daughters of Partik Akhbar, Ovsanna and Mari. "Our mother told us not to return without you," they declared.

The girls reached for his belongings and reeled them close. "Stop," Stepan pleaded. They did not stop. Instead, they propped him up, threaded his arms into the sleeves of his shirt, and slipped his pants over his legs. With Ovsanna on one side and Mari on the other, Stepan was lifted to his feet. By now, he had stopped resisting, not having the strength to fight. The young girls carried the grown man to their tent, where their mother was waiting, just outside. With moist eyes, she ushered her former neighbor into her home, freshly pitched in this desolate field. "Ever since I moved to your street as a young bride, your father became our guardian angel in every way," she said. "There wasn't a day that we were left hungry or without a doctor, and all this without expecting us to reciprocate. So I was always grateful, but there was no way I could show my gratitude. That I might die feeling this debt of gratitude made

my heart ache. We already have one foot in the grave. God has given me
an opportunity at this very last minute to lighten my conscience. Praise
God that He has given me the opportunity to help you."

Inside, she undressed him and lovingly laid him down on a bed already made for him. *So soft,* he thought. A bed with a double *shilte,* the
mattresses spun from cotton and wool, not the stacks of clothes or straw
usually under him. With a covering, she swaddled him. He felt warm.
Now hot. He started to sweat. For days, the perspiration and illness
spilled out, soaking him. On the fourth day, he felt a little different, a
touch better. His diarrhea had ceased, and his energy returned. He wondered what was responsible: the warm bed — or the human warmth?

"*Shnorhagalem,*" Stepan repeated to the Unanian family — thank you,
thank you. He knew what the assistance had meant. "I owe my life
to you."

On the mend, Stepan set his own tent next to Partik Akhbar's, neighbors becoming neighbors once again. With his renewed stamina, he was
back to hauling water but laboring twice as hard as before. Khoren had
fallen ill with the same diarrhea, so each morning, Stepan had to boil
the *chorba* soup before racing to the spring. In time, Khoren improved,
but the interval between each *sevkiyat* had shortened to every few days.
Stepan was worried. Especially since the tension was growing between
the guards and the deportees, many of whom feared the desert beyond
and were refusing to move.

The standoff ended one wintry day when Stepan heard a sudden commotion. A drumbeat of mounted gendarmes had just entered the encampment. His body pulsed, ready to flee. Stepan and Khoren rolled
the tent into a small ball, thrust it under a pile of clothes, and calmly
led their donkey away, as if taking it to pasture. With every step, they
prayed to escape the notice of the guards. Now behind the perimeter, Stepan looked back. The guards had spread out, seemingly everywhere. Despite their commands, some deportees still resisted. The
guards rushed their ramshackle structures and feverishly swung their

batons, beating the young and the old into puddles of bloodied clothes and blankets. Just when Stepan's revulsion had hit its limit, he saw them light some torches and set the tents on fire.

Monsters. Helplessly, Stepan watched the frail continue to lie there, many unable to move, unable to prop their skeletal bodies up, unable to respond to the orders. The ragged clothes and tents and bedding all lit up in a grand blaze that curled to the sky, with the rank odor of burning flesh. Stepan shuddered but was unwilling to look away. His friends, his neighbors, his people were burning.

He could hear their screams. Had this happened a few weeks earlier, when he was paralyzed with sickness, he would have been one of them. "God spared me," he told himself. The masses scooped up their belongings, wrapped them in patterned cloth or *bohchas,* and hustled into line, pell-mell. Some rode in carts, others on donkeys; the majority were on foot.

When it felt safer, Stepan and Khoren returned to camp, now quieted. The sight was unimaginable: In place of innumerable tents, blackened ashes and corpses dusted the field, like a fresh snowfall. The gravediggers were hard at work, dragging the hundreds of bodies and the half-dead into huge holes scooped out of the belly of the earth. Then they began shoveling, the sprinkles of dirt falling in gaping mouths.

Far from this bloodied field, the Gallipoli campaign was finally ending. The Turks defeated the British, Australian, and New Zealand troops, and in December 1915, the Allies withdrew in a daring secretive retreat. All told, there were some half a million casualties, including one hundred thousand men who died in this protracted battle for a strait only a few miles wide. It seemed the Allied forces, which the Armenians dreamed would rescue them, were not coming.

By now, the "Armenian question" was no longer a question. Talaat Pasha, the Ottoman minister of interior, regularly bragged to the American ambassador about his handiwork, "I have accomplished more toward solving the Armenian problem in three months than Abdul

Hamid accomplished in thirty years." So many Armenian bodies clogged a section of the Euphrates River that for one hundred yards, the current briefly changed direction.

Stepan wondered when it would be his turn to leave. By studying the *sevkiyat*, he noticed a pattern to the madness. The people with animals, carriages, and carts were always corralled first. To protect themselves, he and Khoren sold their donkey—the water business was already faltering—and bought some food, which didn't last long. Hunger gnawing, they did any odd job. They bought wood from some Arab women and resold it for campfires. Then the two peddled flour. When that was not enough to survive on, he and Khoren hawked some of their clothes. When that was still not enough, they sold their underwear. That winter, their possessions progressively dwindled down to a few clothing items, bedrolls, and a tent. Enough food for one day, then three more days without.

Each morning, the Arabs streamed into camp, crossing the line of guards surrounding the squalor to set up a bazaar. Some hailed from the outlying villages, others from as far away as Damascus. On some days, the *emanetji* would stand there and salivate. Bread, vegetables, cheese, salt, sugar, and dry wood were stacked a few inches away. When enough change filled his pockets, he purchased bread. Only the rich could afford the prices, items sold for many times their worth. Others loitered too, as if the proximity to food could suppress their hunger, learning the essential words to beg in Arabic: *Dakhílek* (please). Some *khubz* (bread). *Juwaan* (hungry).

Not far away from the Arabs sat a marketplace of clothes, all of them coated in thick dust. Stepan could barely glance at it without succumbing to emotion, this showcase of Armenians' last possessions where he'd sometimes buy clothes himself when he had more money. He saw earrings, rings, undergarments, underwear, bras and slips in all colors and sizes, and fine embroidered silk from women's dowries, all jumbled together. These were their most cherished dresses they had carried, still believing they would wear them for husbands on their wedding days. *Their dreams of a sweet future,* Stepan thought, *traded for bread.* Once well-off, once plump, the women had turned into lurching skeletons, these wasted forms that could've been his sisters.

One afternoon, Stepan noticed an Arab strutting along the rows of tents. He was showing off his new coat, blithely unaware of its feminine cut. Another man also flaunted his purchase, his strong arms looping through the lacy sleeves of a flowing nightgown, his hairy legs awash in silk. More ludicrous ensembles followed, the buyers perhaps unaware of the conventions of Western dress. In another time, in another life, Stepan would have gasped with laughter, but now he couldn't even muster a smile.

As the season ushered in colder temperatures, Stepan and his friends all huddled together and sang psalms. Hearing them, passersby might have thought they were happy, Stepan mused, when in reality, they didn't know if they'd view another sunrise. When the rains began, they continued singing together while the camp squished into mud. Their flimsy shacks shook against the barrage of the wind and the elements. For so long, the earth had been swallowing moisture. Like a sopping-wet sponge at capacity, it could not absorb much more. One night, Stepan awoke suddenly. He could barely see. He could barely breathe. His tent had completely caved in.

Stepan and Khoren shoveled out of their fallen shelter. Flinging away the roof, the walls, the gelatinous slop, they could finally survey the grim scene: Qatma had flooded. A sea of sludge undulated around them, the ground all *chamur,* mud, submerging the tents and the people inside. Each time Stepan took a step, his leg nearly disappeared into the muck, like some wartime amputee.

Shivering, Stepan arranged some kindling into a pile that crackled into a blaze. He felt warmer now; he had nothing else. No clothes to sell, no food, only the night to inspire him on how to survive. *Candles,* he thought. Borrowing forty ghurush from a friend, he was soon pouring wax into small tin molds and then soliciting place to place. The light lifted the gloom, one flicker at a time, in a kind of vigil, allowing him to stock up on essentials again.

Then came another *sevkiyat.* Rushing to the periphery, he watched the guards' rampage. This time, the Unanian family was forced away, their luck run out. Stepan couldn't even say goodbye to the neighbors who had nursed him on his sickbed. He was still reeling when a caravan composed of his townspeople arrived. One man had an oval face, dark narrow eyes. Nerses Aghajanian, Stepan immediately realized, one of his best friends. Who could forget that time they had drunk all that wine? Or the time they put Nerses's parents in barrels on the back of a carriage to save on fare? Several days after reuniting, the old friends ran into each other again. It was good timing, since Nerses Aghajanian had a proposition.

"We're going to Kilis to wash." That was a small town, three hours away, with a bathhouse beyond the reach of the guards. The dirt on their bodies had become unbearable, worth any risk to remove. "Want to go?"

In the quiet of dawn, the five Armenians set out, leaving Khoren behind. The filthy men crossed the outer limits of the camp without detection and then began to scale the northerly mountains. Stepan couldn't wait for a bath, to feel human again, despite knowing it would be fleeting, that he'd have to return to Qatma. In fact, the entire empire had become a prison for the Armenians; there was nowhere truly safe. They came to the ridge's peak, then descended to the plain below, the rocky barrier behind them now.

Plodding eastward, the men encountered a fairly large town, comparable to Adabazar. In Kilis, stone houses rose several stories high, jammed on small alleyways. He didn't see Armenians anywhere. Before the war, twenty-six hundred had resided in this thriving town of more than thirty thousand. Post-deportation, only their schools, houses, and churches remained. In this stillness, Stepan and his friends made their way to a *khan,* a rudimentary inn. The proprietor was sympathetic, allowing them in for forty *paras* apiece. Even though *khans* didn't have much furniture, Stepan luxuriated in the roof overhead.

The next day, avoiding the main drags and authorities, they headed to the hamam. The town had several of the old structures, dating back centuries, square-shaped with grand domes. Making his way there, alongside his old friend, Stepan recalled another night with Nerses at the hamam in Adabazar, when Nerses had mistakenly brought his child's nappies instead of his towel. He could still see his friend in that fantastical diaper turban and hear their laughter together, alongside their three other comrades, the memory choked by a realization: it had been only eight years ago. What a different party of five, very much in need of a bath.

At the Kilis hamam, an attendant blocked the entrance. Stepan's gang of unwashed men lined up anyway. After inspecting them, he declared the price: "Forty *paras* per person."

The group paid together and crossed under the doorway's arch. None of them had a towel; it wasn't just Nerses this time. In the changing room, they started peeling off their soiled clothes. The dirt seemed to be baked into every skin crease, Stepan's long nails bearing crescents of grime.

In the domed area where villagers bathed, Stepan happily noticed a bit of soap. Lathering it up, Stepan delighted in his first real bath in months. He was so hot, he started to sweat. Just as he and his friends began to rinse, he heard someone shouting. The hamam's owner appeared with a stick in hand, his expression livid.

Furiously, he swung it around in all directions, striking Stepan hard. "*Hĭnzĭr gĭaours,* infidel pigs!" the proprietor yelled. "You soiled my hamam."

Stunned, the naked men snatched their clothes and leaped out of the bathhouse, slipping in soap. They sprinted away as the owner tracked them to the town's edge. Finally losing him, the group stopped to dress. Stepan jerked on some clothes and held the rest, and they began to hike the mountains toward Qatma. Coming had been a bad idea, he now knew; Khoren was wise to have stayed behind.

When Stepan reached the camp's perimeter, he saw all sorts of commotion. A passel of Armenians were being separated by profession: the skilled workers were being sent to Damascus, the others toward Deir Zor. Stepan stilled himself until the danger seemed to be past. He knew which way officials would send a man with a third-grade education. Not to Damascus. But he worried, not knowing if Khoren was lost in that crowd, not knowing what the future held.

Afraid, the soapy men hid on the nearby ridge. When they finally returned to Qatma, the field stretched nearly empty. Where thousands of tents had stood, only three hundred remained.

Stepan scanned the piles of refuse and spotted Khoren. Somehow, he'd evaded the raid. He felt relieved but still distressed; fewer people meant fewer customers for candles, water, wood . . . How would they survive? Many of those left had a *vesika*, a paper exempting one from *sevkiyat* and military service, often procured through a *baksheesh*. Stepan and Khoren had neither the papers nor the money for the payoffs. Through the trash heaps, they searched for a discarded container. Finding one, they hiked to the stream and filled it up. Only after much effort did they manage to sell the water. These days, most were fetching their own. In less than a week, the men had nothing left.

Like before, the Arabs were arriving in the early hours, their arms full of goods. On this day, a crowd quickly formed. Stepan lurked around the edge with one solitary thought: *food*. Stepan stared at the *yufkha*, round unleavened bread. He and the others pressed in closer, the wafer-thin vying for the wafer-thin. In the days ahead, the refugees pleaded with the bread vendors to lower prices. Instead, they hiked it to thirty *paras*. One *yufkha* cost what three used to. Famished, Khoren suggested the only option left, "Let's start stealing bread."

And so they did. In Stepan's first theft, they took a bag of bulgur; the wheat grain fed them for two full weeks. No longer a man of practical jokes, Stepan's new circumstances had driven him to desperate new lows.

"Kalkin! Chikin! Yikin!" Another raid on the camp. This time, however, not even Stepan could escape the stampede in the direction of the train station. The gendarmes were clearing the entire camp, supposedly to protect passing troops from the Armenians' diseases. Only the gravediggers were exempt, tasked with burying the bodies left behind. The train didn't seem like it could hold all of them, and yet in the gloaming, the crowds shoved aboard. Stepan and Khoren were crammed inside

too, the trains now operating at twice their capacity. There was little room to move, the men's arms immobilized.

In the twilight, Stepan felt the forward momentum, the fast-moving scenery a blur outside. The cold year was drawing to a close, and any real break in the rain was not expected until March. While Stepan didn't have his family, he still had Khoren and his health. He was better off than many. At the station of Muslimiye, the train halted. They were just north of Aleppo, he learned, and they were transferred onto another line that veered eastward toward Ras al-Ayn and crossed a desert — farther and farther from home. While others slept, he couldn't stop thinking about his family, wondering if he'd ever see them again.

The Bath

2007

THE MATRONLY LADY was chattering away in rapid-fire Turkish. Standing across from her, I could see her face had twisted into a scowl. "I don't speak Turkish," I repeated several times, helplessly. Her voice seemed to grow in force; her rotund body was rooted in the doorway, haloed with ecru and gray stones. She was not letting me into her bathhouse, Pasha Hamamï. I stepped back and gazed at the sprawling structure, which stretched for a block in this town of Kilis. My watch showed two o'clock, the designated time for women. I was done admiring the exterior, with its small grated windows and bulbous dome. My drive from Hasanbeyli had been long, and I wanted a bath. What could be the problem? She quieted now, expecting my reply.

"Jemal!" I shouted to my driver. "What is she saying?"

He had just stepped back into his van, the one Baykar and I had spent the last week in, traveling across southeastern Anatolia. Lately, I had come to think of our Mercedes as the Magical Mystery Bus, with its canary-yellow color and melodies that had now transported us to the edge of Turkey, just five miles from the Syrian border. On our drive one afternoon, Baykar began to sing. For such a quiet man, his voice was powerful and angelic, raising the hair on my arms. Soon we no longer needed a radio; Baykar would sing a verse, and Jemal would join in with old Kurdish and Armenian songs as we road-tripped, the green scenery and tracts of farmland darting past. But now that we'd arrived at the hamam, Jemal and Baykar were already back in the van and about to leave for the afternoon. Over the roar of the engine, Jemal heard me and hopped out. Translating, he said, "Five more. She wants five more lira." I could tell he had pushed back, but her expression was not wavering. In frustration, I paid her another five, the special foreigner price.

Jemal turned to leave. "Are you okay?" he asked, his paternal side showing, as if he were dropping me off at grade school for the first time.

"Yes," I said.

"Call me when finished," he said.

The old woman scanned me again with frustration. She yelled something, and a younger woman scurried into view. She was the *keseji*, attendant.

Now the older one, perhaps in her sixties, waved me to the doorway, as if she were swatting a fly. I took it to mean I should follow the *keseji*. Quickening my step, I entered a large darkened space, the air heavy with moisture. I could almost feel the years, the three and a half centuries that had passed since this place first opened, the warren of rooms derelict from neglect and wear. She led me into a small closet-sized chamber, motioned for me to undress. I removed my Brooks running shoes, my pink short-sleeved shirt, khaki pants, and undergarments. Standing there without a towel, I felt like a dumb foreigner, expecting to find stacks of them folded on shelves. Peering out, I asked for one. The matronly woman sighed and began to forage, eventually unearthing one.

"Shampuan?" the younger woman asked; shampoo. *"Hayr,"* I said in Turkish, one of the few words I knew — no. She breathed deeply, as if inhaling would take her somewhere else. I followed the younger woman into a grand, domed room with a skylight that illuminated the cavernous bathing areas. It was then I realized that both my grandfather and I showed up at Kilis hamams without anything in hand.

A smaller chamber sat off to the side with water faucets and a *tas,* or decorative small bowl, set nearby. The *keseji* placed the *tas* under the rushing water and then poured it onto my head. The water felt warm, refreshing. I could see why my grandfather had felt safe within a bath-house's walls, within its cocoon. She filled up the bowl again, poured it onto my body. Then she repeated the motion before leaving me on my own. Beside me squatted two naked women. They were washing their hair. Observing me fumbling, they laughed and giggled. Shy, I didn't know if I should avert my gaze. I must have looked ridiculous, turn-ing the spigots on and off like a child, an embarrassed smile on my face. One addressed me in Turkish. "I don't understand," I said.

"Where you from?" one asked in broken English.

"America."

"Ohhhhhh," they said.

I hesitated.

"*Ermeni,*" I said, using the Turkish word for Armenian. "Mama Ada-pazari," I said, stringing together my point.

"Ah," they said. They both scooted toward me now, shampoo in hand. One put some in her palm and dabbed it onto my hair. She pressed her fingers down firmly and began washing my hair, swirling it in circular motions, massaging my head. She gathered my long hair at the top of my head, like a bun, and curled the tresses around, as lov-ingly as my mother used to do. I closed my eyes. Then, out of nowhere, they began to sing, just like Baykar had. Their voices echoed, the sound bouncing off the arched ceiling like light.

The melodies were beautiful. I began to clap along, my only musical talent, as we all laughed together. When their singing trailed off, the *ke-seji* escorted me to a nearby platform and directed me to lie on my back. With a loofa, she scrubbed my legs and my arms. The pressure was hard, akin to a contractor blasting off old paint layers. Her hands seemed to scrape away both my dead skin and my misconceptions about the Turk-ish people all at once. In the humid air of the hamam, I felt at peace.

I didn't know what I had expected to find here, but for some rea-son, I intuited its importance. My time in Turkey was nearing an end. It had been different than I'd thought, and I hoped that it reflected the change within the people's hearts. I had just one more important stop to make before crossing into Syria: a visit to the country's last Armenian village, Vakïflï Köyü, the only one from my grandfather's time that remained but where the elderly population was slowly dying off. When I exited the hamam two hours later, Jemal and Baykar weren't there. As I waited, the grumpy proprietor asked, "America?" I replied, "American." Then I corrected myself. *"Ermeni,"* I said, emboldened. She seemed to consider what I was saying and then her mouth lifted slightly in a smile.

Water's Course

1915–1916

OFF THE TRAIN FOR DAYS NOW, Stepan struggled to get his bearings in the desert. From the mountains, his caravan walked southward, the landscape before him flat and endless. In the icy air, he took another step, the line of Armenians long. He couldn't view the Euphrates River yet, but he was headed that way.

Several years before Stepan's march, the famous Baedeker travel guide had accurately warned about the dangers of exploring this region, "An ample supply of provisions should be taken in the carriage, as little or nothing can be obtained en route."

If only he had a carriage — or provisions. It was late December still, and the sentries were escorting his convoy forward, all stepping through dust. A week or so had passed since the caravan had left Qatma by train and then continued on by foot. On this route, a band of robbers were ambushing the Armenians, wringing liras from the lira-less. Before long, the bandits descended on Stepan's caravan while the gendarmes passively observed. In the dimming light, the winding chain of Armenians halted in some hamlet, but more theft needled the night. They took, in addition to the large *bohchas,* the women, especially the beauties, whisking them away like a handful of jewels. Although Stepan's bale was spared, the fear rose upward from his stomach like vomit as he lay there. No sleep for him or anyone else.

The next day, dead deportees studded the path. Many he recognized from Qatma; they were half dressed, their eyes fixed somewhere else, perhaps on a better place. He remembered them from the caravan before his. Some of those who had lagged behind the day before caught up with the caravan, but they had been stripped of their clothes, the bandits' handiwork. Stepan felt blessed to have the shredded garments that still draped from his reedy body, goose-pimpled from the cold.

More walking. In this region, swarms of mud homes rose up, shaped like beehives with compact doorways. What Stepan would trade for a night indoors; his last had been in Kilis, before the ill-fated bath. Human remains marked the way as the convoy curved toward the heart of Mesopotamia. After all day walking, a friend from Stepan's tent couldn't continue and toppled over. There had been twenty-one of them at the last stop, clenched together like a fist.

Some more steps and another friend collapsed. He couldn't mourn them; it was happening too frequently. If he let it, the grief would deplete his strength. At a large clay field, they were told they could stop. The *emanetji* surveyed the new encampment for any business opportunity, but there was none. The town's name was Bab, he learned. *Bab* means "door" or "gate" in Arabic, but he did not know that. He didn't know the language yet, or through which door he had entered.

The temperature dipped lower, freezing the rain into snow. Inside his shared tent, Stepan and the others fought to keep warm. The biting chill seeped in between the openings, the bed of bodies providing the heat. There were fewer of them now, nineteen occupants instead of the previous twenty-one. Unfortunately, the snow had pounded down that winter. Only once every decade or two did it fall like this, locals said. Through gales and clear days, the Armenians tried to hold on to life. The cold slowed the hours and came paired with sickness, with typhus. Stepan now felt hot. Stretched on the ground, he was burning up. Immobile again, just like before. The odds of his surviving weren't good. As the calendar unceremoniously rounded to 1916, the *emanetji* watched more of his tent mates turn quiet, never to stir again.

·   ·   ·

Back in Constantinople, the persecution of Armenians continued to weigh heavily on the American ambassador. Since the deportation began, Henry Morgenthau had been inundated with accounts of the suffering, many of them coming from foreign sources, including Germans, the Ottomans' ally. Morgenthau did what he could to help, but it was not enough. Hundreds of thousands had died, and there was no end in sight. Demoralized, he asked to leave his post. During his farewell meeting with ministers Talaat Pasha and Enver Pasha that January, he raised the issue of the Armenians again. "What's the use of speaking about them?" Talaat Pasha asked, waving his hand in the air. "We are through with them. That's all over."

Not long after, Morgenthau departed the capital, his home for more than two years. The diplomat's train crawled out of the station and cut through a Europe ravaged by combat. Despite the staggering death toll so far, the war had stalemated. Both sides were sandbagged behind nearly the same lines as the previous year. Surviving some bullet fire, Morgenthau's train chugged through a devastated Serbia with its wilted farmland, blasted bridges, and bombed-out buildings, before heading farther north to Berlin.

Morgenthau made his way back to the United States, where President Woodrow Wilson prepared to run for reelection. So far, the country remained neutral. From safely across the Atlantic, the American public read about the war, about the clashes of British and Ottoman troops in Mesopotamia and the Russians' offensive on the Caucasus front. They were pushing west toward Erzerum and Trebizond, once-thriving Armenian towns.

Still haunted by what he had witnessed, Morgenthau penned his memoir. "My failure to stop the destruction of the Armenians had made Turkey for me a place of horror, and I found intolerable my further daily association with men who, however gracious and accommodating and good-natured they might have been to the American Ambassador, were still reeking with the blood of nearly a million human beings," he wrote.

In the camp, the afternoons were indistinguishable until the day Stepan felt arms wrap around him and lift him up. His tent mates were stirring too. More orders, he learned, to transfer to another site in Bab. Still crippled by typhus, he couldn't walk, could scarcely lift his head. It was a good thing he had friends, his sole currency now. One of his friends, he realized, was carrying him.

It was a gray winter existence. He'd separated from Khoren in the last camp so each could find shelter, and now his only company, his tent mates, were falling away. Each day, some four to five hundred perished around him. Remarkably, Stepan recovered, but that January more than a thousand others died within just sixty hours. The shouts of the *mezarji*, the gravediggers, usually Armenian deportees, followed as they called out in search of bodies. Sometimes, no one answered, since all the tent's occupants were in need of their services. The men would drag the corpses by the feet or hair and drop them into an enormous hole; some ninety bodies filled the space. For the removal, the *mezarji* fleeced the deceased's family members of the little money they possessed.

Watching this, Stepan felt disgust. *Betraying their own kind,* he thought. He wondered how the *mezarji* could do this, though he knew the answer: the special treatment from the guards, the extra food, the exclusions from each *sevkiyat*. He felt the same about the Armenian night watchmen. Stepan spied them escorting the sentries around, pointing out the tents of the wealthy so the guards could rape the daughters and steal whatever honor they had left. Those who were unable to afford bribes to ward them off had their tents and their ratty blankets burned. Powerless, Stepan simmered and waited for the madness to end.

Finally in February, a break: the skies cleared, at least temporarily. To stretch his legs, Stepan set off toward the village bazaar, avoiding one body only to come to another. His pulse quickened — hundreds of corpses covered the path. It was a ten-minute trip, but it overwhelmed him, despite all he'd already seen.

Not wanting images of this circulating, Turkish authorities forbade anyone to photograph these scenes. Pictures would surface anyway. Armin T. Wegner, a German second lieutenant serving as an aide in the Ottoman army, smuggled out his photographs of skeletal Armenians in paper-thin clothes, of the distended bellies of the starved and dead. The German concealed the photographic plates in his belt, and the prints became some of the only recorded images of the crime scenes.

At the marketplace, a beggar called Stepan's name. The *emanetji* couldn't place him. A long, knotted beard tangled down his face, his hair as long as a girl's. His shirt and pants gaped open; over the ensemble was a ratty fur that must once have been a fine coat. With great sadness, Stepan finally recognized him from home as the supplier of clothes when he was peddler. Now the *efendi* asked for bread. He needed help; all Armenians did.

That month, approximately five hundred thousand Armenians wasted away in the triangle between Aleppo, Damascus, and Deir Zor. The mounting evidence of an annihilation prompted the American secretary of state, Robert Lansing, to write a sharply worded warning to the German ambassador. A few months earlier, the Germans had promised to assist the Armenians, yet they had not. Now Lansing accused them of not doing anything to "prevent the repetition of excesses against the Armenians."

Nothing changed. On one of these dreadful Bab days, someone saw something in the sky. Was that an aircraft? Yes! The war has ended! An armistice has been signed! This spread through the camp like one of their sicknesses, but this time boosting hope. Really, though, the light above had been bright stars, the first onlooker twisting a wish into a truth. In fact, the war was far from over. More blood would flow for another two and a half years, and in Anatolia, it would continue until 1923.

The sentries charged into Bab one March morning, signaling yet another move. Stepan joined the ruckus of pleas to stay, all for nothing. Without hope, the deportees with camels assembled first, securing their belongings atop the animals. Stepan rolled up his grimy blankets and rags and took his place, the sick trailing the rear, the line swaying out of camp like a snake. Stepan scanned it. *Unbelievable,* he thought; the head was positioned an hour's distance from the tail.

The mass exodus was due to a new Interior Ministry directive. All Armenians in their *vilayet* had to move on. Despite almost a year of death marches and massacres, many Armenians remained stubbornly alive. The objective now was to close the encampments near Aleppo and push the outcasts farther into largely uninhabited terrain. To prepare for this final stage, Talaat Pasha reorganized the leadership in provinces and had the Armenian orphanages cleared. This marked the beginning of, as one historian described it, the holocaust's "Second Phase," which followed the first wave of deportation.

The net widened to include more Greek Orthodox Christians, ousted from their homes in Alexandretta and Antioch and other locales along the coast. By now, many of the previously excluded Armenians had been deported too. Many exiles were driven along the Euphrates, where camps were closing from west to east, forcing the Armenians and other Christians closer to Deir Zor. Around the same time, the camp's *mutasarrif,* a humane man, received orders to expel thirteen thousand because of the wartime rule prohibiting Armenians from exceeding 10 percent of the local Muslim population. But send them where? For great distances, there was nothing beyond Deir Zor, everyone knew. The *mutasarrif,* Ali Suad Bey, felt for the deportees and had been helping them. For that, party leadership had singled him out. His days at his post were now numbered.

As the Armenian deaths continued to rise, the Russians advanced farther into Anatolia on the Caucasus front. Driving the Turks back, they captured Erzerum, once flourishing with Armenians. The emptiness was haunting — almost no one remained.

For this, the Russians exercised rough justice on many innocent Turks. Stepan didn't believe in revenge. During all those days of marching endlessly, he increasingly looked to his faith for answers over the murder of his people. *There is a higher law,* he told himself. *God will adjudicate the deaths.*

Fighting exhaustion and thirst, Stepan's convoy slogged forward as a host of foreign villages tumbled past their feet. At one stop, mothers sold their daughters to Arabs for a handful of ghurush. Others gave them away for free. Eight-year-olds or eighteen-years-olds, it didn't matter; this seemed to be the children's only chance at life. The path was clotted with bodies sprawled on their backs, and more from Stepan's caravan added to the grisly procession. Stepan recognized many from Bab. One deportee on this route passed the miles by counting the dead. In one hour, the figure rose to five hundred, mostly girls, many partially consumed by wild dogs. Up ahead, some water, some hope. Stepan could see it too — the Euphrates.

Stepan reached the edge of the river and stopped. The waves were churning, listing to the side. In this area, the southbound river begins to bend like an elbow and course eastward. He had read about "the great river" in the Bible during all those Sundays at church. It was the lifeblood of a land bereft of water, and because of this, it possessed a divine-like power. Many Armenians hoped it would answer their prayers for deliverance. Some merrily drank from its cup while others jumped in holding their children's hands — and did not come up again.

So many ancient societies had flourished, and withered, around its banks. Mesopotamia, the cradle of civilization, an area Stepan had always heard about but never imagined visiting, so far from home. This was the land "between two rivers," the land that spread from the Euphrates to the Tigris.

The riverside town of Meskene stood at Stepan's feet. It had been forty-five long miles since Bab. Out in the field was another wretched camp, filled with an estimated ten thousand denizens, countless more deceased, all within view from the road. The worst so far, Stepan decided, worse than Qatma. His view was prescient; in the year to come, the number of dead would swell to sixty thousand.

Stepan was winded, but the caravan didn't halt. Instead, the charge guard ordered them onto a path that shadowed the Euphrates. Stepan began to track the river eastward. Past Deir Zor, the waterway turned toward the equator again and joined the Tigris River, the two eventually emptying into the Persian Gulf. All that after such humble origins in northwestern Anatolia, the historic home of Armenians. The

Bible had placed the Garden of Eden near these waters and even pre-
dicted that the day the Euphrates dried up would signal Armageddon.
But Stepan didn't need to wait for such a sign — it already felt like the
end of the world.

Now, on the Euphrates' right bank, Stepan marched onward. Beside
the convoy, the guards rode on horseback, howling orders. The trail cut
across a gypsum plateau, and a long ridge of terraced rock chased alongside them. Over his other shoulder, Stepan might have noticed small
islands freckling the broad stream. It had to be comforting to be close
to such a large supply of fresh water, the border green, jungle-like with
shrubs. Two hours eastward and, suddenly, a stop. The route was impossible for carriages. So the long line of people switched direction and
scaled the bluffs, continuing on the cracked surface of mineral deposits,
navigating their step as the light fell away. And still they walked.

Stepan knew little of the war's developments. His enemy was each
new day on the road. The military traffic along the Euphrates was
heavy, with troops heading toward the Mesopotamian fronts. In Kut al-
Amara, the Ottoman army had surrounded the invading British forces.
Without enough men, the British conceded to the Ottoman army;
nine thousand soldiers were taken prisoner. Far away, on the western
front, the Verdun offensive claimed more lives as the Germans fought
in northeastern France, the battle growing into the longest of the war,
with a German commander promising to "bleed France white." He succeeded in bleeding his own army white at the same time. Farther north
on the river Somme, British and French forces attacked the Germans.
Both battles resulted in further stalemates, with casualties for 1916 on
the western front approaching two million men.

At sea that spring, it was no safer. A German submarine mistakenly
hit the *Sussex*, a French steamer full of civilians. Though all survived,
twenty-five onboard were Americans. In response, the United States
threatened to sever diplomatic ties with Germany. To prevent that, the
Germans made their "Sussex Pledge," promising to warn merchant vessels before an attack and assure the safety of passengers and crew. This
kept the Americans out of the war — for the time being.

A large valley spread before Stepan, choked with two-thousand-plus tents around that time, many of the Armenians gravely ill. The village was called Dibsi; it was where Alexander the Great had famously forded the Euphrates. Now the place was known to the deportees simply as the Hastahane, the Hospital, except there was no medicine or healers, only patients. By peering at the gravedigger's ledger, Stepan gleaned the seemingly impossible figure: ten thousand had died in this one camp in three months.

That April, the guards escalated the *sevkiyat* by burning down the tents. When Stepan's erupted into flames, he scrambled out and used the last of his money, forty ghurush, to load his bale onto a rented horse. All day he walked, and just before sundown the village of Abuharar appeared. Soon Stepan would sleep. As he retrieved his belongings from the horse, the animal's owner stopped him. "Sixty ghurush," he decreed.

"We had agreed to forty," Stepan said.

"On the way, I had to give twenty ghurush to the gendarme."

A *bekchibashi* overheard the dispute and solved the problem by walloping Stepan. He screamed, but the beating did not stop and wouldn't until he produced the cash that he did not possess. Out of options, he did what he always did in these situations — he sold more of his cherished garments — before being forced onto the road again to yet another hamlet.

Often, Stepan's caravan crisscrossed soldiers either looping back from the Baghdad region or deployed there anew. Several months earlier, one of the carriages on Stepan's road had held the German consul of Tabriz, Wilhelm Litten, who gazed out at the passing scenery. It was a rare view into a forbidden territory. "I have seen with my own eyes about 100 bodies and almost just as many fresh graves on the road from Deir Zor to Meskene," he wrote. "I have not counted the graves which in some towns were combined to form cemeteries. I have seen around 20,000 Armenians. I have restricted all my estimations of numbers to those I have actually seen for myself." Litten sent this dispatch to Walter Rössler, the German consul in Aleppo, along with a time-stamped registry of the stretch where Stepan now walked. The consul spoke of all the bleached skeletons, the faces black like masks. The sleeping beauty with the missing right arm. The dead child, and the pudgy dog ready to pounce. He passed them all, then he came to this:

4.04 hrs: 1 skeleton on the road, close to the wheels of the carriage. Teeth and flesh on lower half of the face still existent. Facial expression therefore a broad smile over bared teeth. A frightening sight. [On the left]: on a small rise, therefore roughly on a level with my eyes, a female child of about two years, only clothed with a red bodice which is pulled up. Bleeding genitals revealed and facing the street.

As the weather warmed, blades of grass appeared on the riverbanks, sprouting like a baby's first hairs. Stepan picked some, lifted it to his mouth, and began to chew, his only meal now. On more notable occasions in this somber place, he sold another garment and bought a little flour to dredge the grass in before cooking.

It was there he'd heard that another of Adabazar's finest was dying. Visiting the tent, Stepan saw a husband, wife, two kids, their eyes and cheeks sunken. Thousands of insects were crawling on top of them. Stepan thought he'd come too late. Then the man spoke. "My friend, please give me some water for the last time." With his narrow finger, he was pointing to a rusty tin box beside the bed. Still startled, Stepan looked in: there was not a drop left. He wanted to help him with something more than water, but he, too, had nothing. He filled up the container, set it within reach of the emaciated man, and fought back his sobs.

Twenty days or so passed, and then came an order to leave. On his way to the road, he encountered some people strewn on the ground. Three skeletal children hovered over one of the recently deceased. *"Mayrig, Maaay-rriiig"* ("Mother, Mother"), they cried, tears drowning their eyes and cheeks. With a heaviness, Stepan looked at the dead woman's face and recognized her as another of his hometown's wealthiest. One by one, they were all disappearing, all his neighbors, friends, and community leaders. What future would be left?

The convoy stopped in yet another clearing with thousands of tents. Under Stepan's roof huddled just three now: himself, a woman, and her young cousin. In the past six weeks, the other eighteen had died, including the father who had taken pity on Stepan and invited him inside. Hunger now burned in the three of them in this village named Hamam, making them forget all else. They fell asleep hungry and woke up hungry; they sold their paltry belongings and traded the sum for one loaf. Greedily, they fell upon it, the problem for the next meal renewed. For months, Stepan hadn't eaten properly, but lately he couldn't cope. He dreamed of the sensation of having something between his teeth. April 15 came and went, and Stepan marked his thirtieth birthday, owning little more than what he'd had when he entered the world. Easter was fast approaching, his second one away from the family that he thought of each day. There was no need to fast this year; he already was. A few days before the holiday, his tiniest tent mate stepped to the river's edge. While fetching water, the young girl slipped; her small body was carried away by the current, reducing the final group to two. He was in disbelief. To survive this long and then drown accidentally? In the wake of her death, a relative of the two cousins rolled up the tent and carried it away, like some inheritance of the damned.

Now Stepan had neither food nor shelter. For hours, he observed his comrades on their knees, scouring the ground for thorns to sell or scrutinizing the behind of a donkey as if it were a film — lying in wait for the animal to defecate, ready to hustle over and comb through the dung for an undigested barley seed or anything else. *Eh, this is life.* Stepan roamed in search of work. There was none. He searched for bread. There was none. Sometimes an acquaintance from Adabazar who was also in the camp, Hovhannes Bekhurian, graciously offered him a slice.

When a dead dog washed up on the banks, it seemed like dream. This great river had finally gifted the people some food. A group charged and carved up the furry carcass, ravenously lifting chunks to their mouths. Then, by some luck, a donkey keeled over a little distance away. The strongest ones journeyed and butchered it; the meat lasted for some time.

The whirl of energy and happiness gave the camp a festive air, much needed as rumors spread about what a mother had done to her own child. Soon another, similar report surfaced. *Gossip*, Stepan thought, consoling himself. Then he heard of it again and investigated for himself.

In a circle sat four girls and a woman. Evidently, it was mealtime, as they reached into the center dish and sliced their portions with an old jackknife. The body of the elderly female lay there, mutilated. They had been at this for quite a while. A lot of the flesh had been carved away, up to her hips, scraped out. The girls now ripped the meat off the ribs, the remaining cuts divided up like a cow's. A heave of revulsion rose up.

"Do you know what you're doing?" he blurted out.

The women looked up at the stranger who stood there, judging.

"*Akhbar*," one said; it meant "brother." "What can we do to not die of hunger? In the old days in our villages, when we saw a dead person we were scared, but today we are in this situation."

After the inquisition, the women returned to their meal. One by one, they cooked their portion, boiling or charring or smoking the slices of human leg, then setting it down on galvanized plates. As they forked the flesh into their mouths, Stepan left. In the days ahead, he would witness cannibalism innumerable times and steadily become inured to its horror.

The hunger stretched on, almost with an appetite of its own. Stepan passed the unrelenting hours by lying on his side, gazing up to the sky. He had gone a good distance from the tents to do this and was now deeply lost in thought, contemplating all that he had seen. *Eh, I've survived another day. Now what?* Staring up, he viewed the sky some more.

"*Akhbar*."

Stepan thought he heard a voice.

"*Akhbar*."

It seemed clear now. Spinning around, he discovered his bread benefactor, Hovhannes Bekhurian, standing before him.

"Where have you been?" Hovhannes asked, and he seemed to be in an especially cheerful mood. "I've been looking for you for an hour. Come, I've found work for you. I felt bad to see you starving all the time."

With that, Stepan felt himself lifted out of this dark place, as if he'd been rescued from the bottom of a dank well. Eyes wide, he sprang up, renewed. Apparently, the Adabazartsi Garabed had hired someone from the local villages to drive his cart, which he managed to keep somehow. Naturally, Hovhannes confronted the cart owner about this: "Wouldn't it be shameful if our compatriot Stepan, a respectable young man, died of hunger before our eyes? Why don't you give him a job with your cart?"

The man seemed to consider this but remained wary.

"It's a pity, my friend. You'd save a life."

Garabed relented, and Stepan soon found himself transporting deportees of means during each *sevkiyat*. Somehow, the shadow business was allowed to persist, which hastened the sweeping of the Armenians farther east. Stepan and other drivers were steering the refugees and their goods from place to place, especially the short distance between Hamam and the edge of Raqqa, a medieval town on the left bank of the Euphrates. As the spring months fell away, the *emanetji* steadily got used to the tug of the two oxen in front of him — and the joy of daily food, thanks to his payment of hard bread. During one of his transfers, though, the vehicle jerked to the side and then careened to a halt; the wheel had splintered into pieces. He needed to get it fixed. The nearest repairman was in Raqqa, which he'd never actually set foot in before. Not only was there a severe penalty for entering the municipality without registration, but it straddled the other side of the river, and there was no bridge across. Many Armenians longed for admission, having heard that some Arabs were sheltering Armenians in their homes. As a result, Raqqa was swollen with refugees — eight thousand, the lucky ones able to make it inside the municipal limits, a rare safe haven along the Euphrates.

At the great river, boats ferried people back and forth, forming an expanding V in their wake, the traffic often heavy with German barges and convoys loaded with cannon and artillery. Stepan needed a ride across. But how? The *sevkiyatchi bashi,* chief of the expedition, was scrutinizing passengers. Eyeing the strong current, Stepan wondered if he could swim it, the fate of the drowned girl fresh in his mind. Nervously, he approached the *sevkiyatchi bashi* and explained the situation. The feared official listened and stared. Then he granted Stepan permission to go.

The barge seesawed as it crossed the inky surface to the other bank, where the *emanetji* disembarked with his broken cart. Amid the town's houses of crimson brick, he found his way to a repairman. While waiting for the wheel to be fixed, he ran into Khoren Mkhjian, his dear friend from home, his battalion comrade and water-delivery partner. The two greeted each other with much emotion. Khoren explained that he had taken refuge in the town, and he beseeched Stepan to do the same: "Stay in Raqqa. It's safe."

Stepan didn't have to deliberate. While no place was truly immune to the deportations, he had a chance here, especially with his friend and partner nearby. Now he needed to inform his family of his new whereabouts, and he set off to the post office. He hoped they were still alive. The last he'd heard, they were in Bolvadin. Other Armenians tried to keep in touch with separated family members, and many sent money through the mail. Some five hundred to a thousand pounds arrived each week in Raqqa, but the postal employees made it difficult for families to collect the lifesaving funds.

Stepan had so many things to say in this letter, but he distilled it to basics. Curtly, he wrote in his impeccable script, "Send me some money so I can make a living as a street vendor." He addressed it to his brother, Armenag. Return correspondence would most likely take four or five days. Ample time, he thought, to give the cart back to its owner in Hamam; he didn't want to steal from the man who had saved him. He bade his friend goodbye and, with his newly fixed cart, boated back over the Euphrates.

For three days, he bivouacked in Hamam, a barren place. The locusts had consumed all the precious grass, so the people were now eating the insects instead. With the rising temperatures, the clouds this time of year disappeared like ghosts. He'd stayed longer than he'd planned — and now it was too late. A retinue of gendarmes exploded into camp, clearing the tattered tents.

Unlike before, there were no exceptions. Even the cart drivers had to go. Taking the reins, Stepan navigated the road east, his boss's family piled into the cart's back. When Raqqa's skyline came into view, his eyes fixed on the town across the river, and he yearned to be there. He knew that if his family was still alive, there could be money waiting for him. *Any moment, we will stop. Any moment . . .* In front of him, the dispossessed shuffled forward. With each step, his anxiety rose.

Now the caravan was past Raqqa, and his opportunity was gone. *My family will be thrown into consternation . . . wondering what happened to me.* He worried, too, about Khoren. *I should have stayed in Raqqa.*

All day, the convoy advanced, and their numbers further dwindled as many more died, especially those who had eaten the taboo human meat. He could always tell their diets from the way their bodies swelled.

His despair rooted ever deeper in the next village of Sabkha. The deportees spent that entire June in a sun-beaten field wishing they were home. And then something astonishing happened. Two mounted officials shouted to the crowds of starving Armenians, "How many are you? Where are you from?"

It was a census, Stepan understood, as he gave his name and hometown. After they were counted, the administrator shared a bit of news.

"You are going to be sent back. It has been decided to return you to your villages."

Stepan rejoiced; they all did. The camp's sorrow lifted. *I'm going home. I'll see my family.* For the first time in more than a year, he felt free, unshackled. When he had occasion to smile, and this was one indeed, his thin lips curled upward without a showing of teeth, and his cheeks rounded to apples. For so long, he had waited to hear this, enduring what he was certain was every circumstance imaginable. Exuberant, he returned to his tent and waited. Day cycled into night and back again.

During that summer of 1916, the Arab Revolt had begun, with Sharif Hussein, the emir of Mecca, declaring his independence from the Ottomans. For centuries, the diverse peoples of the Ottoman Empire had been governed by the Turks, the majority united by a shared Islamic faith. But with the rise of Turkish nationalism and the long-standing discrimination against the Arabs, many factions dreamed of independence.

With a bouquet of promises, the British had successfully lured the Arabs into their Mesopotamian campaign. Maybe this coalition would swing the conflict toward the Allies and end the Armenian nightmare.

In the middle of July, the wait seemed to be over for Stepan and the others. A line of gendarmes converged in Sabkha, along with armed sheikhs (local tribal or religious leaders), and a guard addressed the cowering crowd, "You soon will be sent through Deir Zor. [Then] get on the train in Ras al-Ayn." Stepan knew that from that station, he could, theoretically at least, ride west to Aleppo and transfer to different trains that would take him to within a few miles of home. He had no idea that Ras al-Ayn had become a slaughterhouse where tens of thousands of Armenians had lost their lives. Two hours later, Stepan heard the usual command.

> Walking and walking, my legs were unable to move,
> Crying and crying, my eyes were unable to see,
> Oh mother! Oh mother! Our condition was too lamentable,
> At the time we were in the desert of Deir Zor.

A hymn from the road. From caravan to caravan, many of the Armenians sang the identical melodies, the majority of the verses eerily the same. This last stretch to Deir Zor would take Stepan and his convoy of one thousand people four days. All the while, he could feel the summer heat beating down on every inch of him as the guards prodded him forward along the river. He'd always remember this path. The water does not forget its course, as the Armenian proverb said. Now he walked beside many other Adabazartsis — he would stay with them until the bitter end. Their names would become imprinted on his memory: Antranig Giumiushian with his wife and daughters and son-in-law Garbis Aharonian. The bespectacled educator Harutiun

Atanasian with his beloved wife, Lusi. He was the one who, six years earlier, had printed Stepan's advertisement announcing his new services as an *emanetji*. He was also the one who, when the Armenians were imprisoned in the town's church, had betrayed them and directed the Turks to the location of some old weapons.

On this day, the thirty-two-year-old Harutiun and his wife marched the same road of thirst as the others. Struggling forward also was Bedros Dimijian, a man who had ventured far from home at least once before. People referred to him as Hajï Bedros in honor of his pilgrimage to Jerusalem. Hajï Bedros was with his wife and sister-in-law, Vartanush Fenerjian. There was also another Bedros, known as Bedros the foreman, and he too was accompanied by his family. Together, they made up Stepan's doomed caravan.

Ma'din. *Forty-four miles to Deir Zor.* This was the convoy's next *otevan*, or resting place. Stepan and his procession stumbled on, three days without a bite to eat. Bedros the foreman dragged himself a good number of paces but couldn't continue and lacked the money for a donkey to carry him. He fell onto his side, alive but immobile. Bedros's wife and daughter set down a bowl of water and kept going. They were not permitted to slow, and the guards made sure they didn't. Up ahead were the crumbling walls of Halabiye. Just across the river, like some type of architectural twin, was the site of Zelebiye. Made of gypsum blocks, these sixth-century cities were the vestiges of the Roman emperor Justinian and still stood in remarkable condition. Onward Stepan pushed. It was on this expanse, wrote the German second lieutenant Wilhelm Litten, where the "horrific parade of bodies" began, as did the ghoulish sights: the thirty-nine freshly dug graves; the old campfire and its ring of six dead men; the body of a young blonde, her arms spread out as if welcoming the sun; and dog after dog devouring yet another cadaver.

In his journal, he noted with icy precision:

2.25 hrs: [On the left] by the road: 1 woman, lying on her back, the upper part of her body covered by a shawl wrapped around her shoulders, lower half eaten away, only bloody thigh bones protrude from the shawl.

Tibni. *Twenty-eight miles to Deir Zor.* Another *otevan.* It was late at night when Stepan's convoy arrived. The exhausted group joined the cluster of tents positioned between the rock terraces and the river. Some wealthy Adabazartsi women were always staggering around the encampment. Often he watched them move. Images of them back home flashed into his mind. How they wouldn't deign to eat bread, preferring fruits and meats instead. How they used to debut a new outfit once a month, frequently custom-made and hand-stitched. How he used to envy them. Now, on bare feet, they begged from tent to tent for a bite of bread. He had none to give, but he wanted to give. The command sounded again. They all rose to their aching feet and marched to the road again, the final leg to Deir Zor.

The Dead Zone

2007

AT THE BUSY BORDER CROSSING INTO SYRIA, the sun beat down onto the stream of cars that twisted out of view. A thousand reflections scissored off the windshields, mirrors, and hubcaps on vehicles that packed the road like an old junkyard. Momentarily blinded, I averted my eyes and turned toward the forbidding checkpoint out of Reyhanli, Turkey. I was nervous, jumpy at every horn blare. Only fifteen minutes had passed since Jemal had dropped me off to park the van, but it seemed he'd been gone too long. In another fifty yards, I'd face the uniformed guards, and I had to get my story straight. I glanced at my security blanket, my gray cell phone; approximately ten in the morning, and no missed calls. In the swell of heat and exhaust on this August Friday, I felt like collapsing. On the asphalt, massive cargo trucks leaving Turkey idled. They roared on, advanced a few feet, then went off again, while pedestrians zigzagged through without fear.

I was keeping an eye out for Jemal, but my gaze kept returning to the surrounding mountains, parched and imposing, the color of dirt and not much else. The place couldn't be more different than where we had been the afternoon before: relaxing high above the clouds in Turkey's last Armenian village, Vakïflï Köyü, a serene oasis of stone homes and tall trees that perfectly followed the sweet serenade of the Kilis bathhouse. I'd been healthy there, too, but had fallen sick later that night. Now dehydrated and dizzy from repeatedly throwing up, I took another swig from my plastic bottle. *Water.* It was all my body wanted; I could never wash enough down my sore throat. I studied the rises from ground to peak. *Could I hike across these heights as my grandfather did? In this condition?* Somewhere beyond this point, he had suffered the most.

I can turn around now, I told myself. But I could not. I had spent too long getting to this corridor that once formed part of the Ottoman Empire.

Only halfway into my nine-hundred-mile journey, and I was already weary. And I was a well-fed thirty-six-year-old traveling by car and train, not on foot as Stepan did. I was the one sleeping in beds rated by stars, not outside on the hard ground under the constellations. Just an hour after leaving my air-conditioned hotel room, I was weak and feverish and needed a bathroom. And I was still far from my endpoint, a godforsaken mound of dirt named Marqada just short of the Iraqi border, where my grandfather's caravan of thousands met its end.

I glanced at my cell phone again; only five minutes had elapsed. I wasn't even moving, but rivulets of sweat were nose-diving off my forehead. Not only did I look ridiculous in my mock neo-colonial disguise of a hat, sunglasses, and baggy linens, but now, I feared, I appeared suspicious too, as if I were hiding contraband. *Where is Jemal?* Passport control would ask why I'd come to the region, and I wanted him there in case anything went wrong. I hadn't been completely up-front about my research when I'd applied for a tourist visa to both countries, and that incident with the Turkish police in Bolvadin was still at the forefront of my mind. I wondered if there was some sort of file on me.

At last I heard Jemal's cheerful voice: "Hello, Dawn!" He was carrying my luggage. With the crowds, he had had to park a distance away. After a week of his guiding me over the mountain ranges my grandfather had been forced to climb, I now had to say goodbye. I would miss his jovial laugh and his protective paternal instinct. In a few minutes, I would cross the border, the portal to the Euphrates camps, Deir Zor, the Khabur River, everything I'd read about.

Escorting me to the guard's booth, Jemal asked for my passport. I extracted it from my money belt, always embarrassing, as I wondered what others must think of us Americans, holding our valuables so close to our underwear. He handed over my blue identification card, and the two began to talk. I smiled to deflect my nervousness. I should have been watching them, rehearsing my lines one last time, but my attention retreated to the mountains. *Could I climb them? No, I couldn't. Certainly not today.* How did he do it? How did anyone?

As the officer questioned me, I tried to answer like any tourist would, expounding on my lifelong desire to visit the ancient riches of Turkey and Syria, which was partly true. I wasn't lying per se, and I wanted to be more forthcoming, but given that it was illegal to say the words *Armenian genocide* in Turkey, I didn't want to take any chances. The Turkish guard waved me through.

"*Teşekkür ederim.*" I thanked Jemal with my few words of Turkish.

"*Rica ederim.*" He smiled broadly.

As I shouldered my luggage and began to walk away, my legs felt unsteady beneath me. One more hurdle to go, with the Syrian guards. I was so close now, and I thought about the kind Arab sheikh who had sheltered my grandfather, saving his life, and whose family I was traveling to Syria to find. Not even my mother, Anahid, knew about this part of my quest. At seventy-nine, she was already beside herself with worry about me, her only daughter, and this leg of the trip made her the most anxious.

In many ways, Syria was a police state, intolerant of dissent and willing to imprison those who threatened the regime. The country was ruled with a firm hand by President Bashar al-Assad. I'd heard about his infamous Mukhabarat, commonly called the Syrian secret police, which tracked its citizens and routinely accused foreigners of being spies. Still, I hadn't been that concerned until a family friend from the region warned me about traveling there alone. "You don't know Syria," she'd said. "It's not like other countries." The United States warned travelers against going to Syria and accused Syria of sponsoring terrorism. Just the year before, armed men had attacked the American embassy in Damascus with grenades and explosives, killing one person and injuring thirteen.

"Dawn! Dawn!" I heard Jemal call out. "You . . . cannot . . . walk . . . border," he cried in his limited English.

"Yes, I can."

"No, it's too far," he said.

"How far?"

"Too far walk."

"How far?"

"Maybe five kilometers?"

"Oh." I glanced at the road that soon disappeared, exposed to the glare of the sun, without any visible shade. Before coming, I had imagined a short footbridge, like the one from California into Tijuana, Mexico. Instead, there was a vast no man's land I'd have to cross. I was so stupid. After a few paces, I was melting.

"Can you take me?"

"I'm sorry, Dawn, no have papers," he said.

Within moments, Jemal went to work on a new plan. "Wait here." Striding into the middle of the road, where the bottleneck lifted, he began waving to passing vehicles. *What is he doing? He's going to get run over.* Several drivers slowed, exchanged words with him, and then accelerated. Finally, the door of a station wagon swung open, and a middle-aged guy popped out. He and Jemal approached me.

"You give five dollars," Jemal said. "He take you."

The man studied me, waiting for my reaction. He wasn't much older than me, short, with dark hair. Shifting his feet, he seemed to be in a hurry. Without another choice, I reluctantly dug into my bag, and withdrew five U.S. dollars, crumpled among my Turkish liras.

Now the man gestured me to the back passenger seat. Ducking in, I realized it was overloaded with men. Not a single woman. Two sat in the front seat, three in the space next to me, and a pile of others in the trunk area with the baggage, knees pressed to chests.

"Hello," I said. Nothing. I tried a few words in Turkish. They still didn't respond, continuing their chatter. They spoke Arabic, I realized, stupidly slow through my fog of illness. Thumbing through my phrasebook, I attempted the basic greeting: *Marhaba, shlonîk?* Hello, how are you? I mangled it so much their eyebrows furrowed. Inside was stuffy, the vinyl seat sticky with sweat. I knew nothing about them; they might have had contraband in the back.

As the car accelerated, I stared out the window, getting a closer look at these desecrated mountains that formed the barrier between Turkey and Syria. I had imagined the landscape of this area would be desert, similar to Palm Springs. But somehow it seemed more forsaken. It was near this border that my grandfather had first realized his life was in danger, that a much darker objective was driving the deportations.

The station wagon sped forward until a Turkish guard blocked the road. Another checkpoint. Apparently, we were still in Turkish domain. The guard scrutinized me for a little while, then spoke to the driver. He handed over his paperwork. The official flipped through the pages, then pointed to the side of the road. We were being pulled over. I panicked. Had the guards at the border crossing figured something out about me? A queasy feeling welled up in my throat. I heaved. I was about to vomit again.

The car turned to the right, parking opposite a small office. I took out my cell phone to tell my fixer on the Syrian side what was happening. *God.* I tried to dial. Nothing. *Goddamn it.* I tried again. I glanced down — there was no signal, and I was on my own.

Hell

1916

> The birds flew away from the trees,
> My heart is on fire, blazing,
> Don't burn, my heart, don't burn,
> This separation was our fate,
> This emigration was our fate,
> This Deir Zorlik was our fate.
> — *A popular song from the caravans*

AT FIRST, Stepan could see only the limits of a town. As the caravan lumbered closer, a wash of white buildings rose up. Deir Zor. Through the plumes of dust, gendarmes now materialized. He braced himself. *Another handoff.* For months, he and at least a thousand others had been on foot, moving toward this place. Soon they came to a bridge over the Euphrates, where more sentries blocked both sides. "Cross," one ordered.

Above the water, a frightening view opened up, a macabre metropolis of tents and Armenians spilling outward from the riverbank, in nearly every direction, onto a plain of shrubs and dirt. The largest yet, baking at high temperatures under a desert sun. Stepan passed skeletal dwellings filled with skeletal people, their waxen arms stretched out, begging for food. Multiple generations stuffed under crudely assembled tents made of blankets, shirts, any material. He skirted around the edges as the guards guided the newcomers, forcing them into a small clearing, where they'd settle for the night, week, month, however long — Stepan did not know.

"As on the gates of 'Hell' of Dante, the following should be written at the entrance of these accursed encampments: 'You who enter, leave all hope behind,'" wrote one witness in a letter to Jesse B. Jackson, the American consul in Aleppo.

The next day while exploring, Stepan spotted a wall of guards, their eyes like hawks', stationed around the periphery some fifteen minutes away. *It's an open-air prison,* he thought, with some sixty thousand inmates. Unlike Qatma, there was no way out. At all hours, the smell nauseated; the urine, feces, all turned to rot in the heat, as the filthy Armenians cooked outside, the high temperatures prepping water for hot tea, stones into oven bricks. Each day, hundreds more succumbed, their bodies interred in mass graves. The *emanetji* was newly horrified, all the more so because he believed he'd already seen the worst. Settling into his new camp routine, Stepan located the thicket of Adabazartsi tents. At the moment, everyone was abuzz with chatter about a large group that had just left. No one could say where they'd gone; they could only gesture to the wasteland within view, seemingly hostile to human life. Yet more convoys arrived. To Stepan, this upheaval felt different than the others — the Armenians seemed stockpiled.

At every turn, Stepan fought not to lose hope, but it was getting hard. He tried to subsist with his earnings from Garabed's cart. But there were few loads or passengers. No loads meant less money for his boss, which meant less *lavash* for him. Still, he tried; only handouts were sustaining him now. The hunger pangs, which had quieted since Hamam, roared again, his misfortune compounded one afternoon when another of the cart's wheels broke. He had to get the wheel fixed. He made his case at the bridge, and within moments, he'd reached the other bank with his *vesika*, permission slip. This Deir Zor was bigger than he'd anticipated, the largest town for more than five hundred miles, midway between Aleppo and Baghdad. On the waterfront towered a mosque, and a bit inland huddled European-style buildings, churches, and more mosques. He walked down the paved streets. Drifting around, Stepan saw people in the routine of life in their customary flowing dress. He must have stood out. His rags were perennially tattered, as if someone had shredded them with a knife.

While waiting for the repair, a loud commotion startled him. He could make out gendarmes rushing into the streets, storming the buildings to uncover the frightened Armenians in their hiding places. They'd been the lucky ones, finding refuge there, away from the camp. No more. The new governor threatened to hang anyone sheltering an Armenian.

Stepan clutched his *vesíka* tightly. It was time to go. He trundled the fixed wheel toward the river. The weather was so hot, he thought he could evaporate. Shortly before the overpass, he planted himself in front of a bush to rest. In the shade, a rustle sounded. Leaves crackled and crunched. He whipped his head around and spotted a man hidden in the vegetation. *He's Armenian,* Stepan thought, the hunch confirmed in the next moment. *"Ov es tun?"* the man asked in their shared tongue. Who are you?

"My name is Stepan. I'm from Adabazar."

The man didn't respond; instead, he disappeared. A few minutes later, another figure took his place. "Apparently we're from the same town," he said, and he asked what Stepan did.

"I was a traveling peddler in our town," Stepan answered. "After the constitution I was a *emanetji* going from Adabazar to Constantinople and from Constantinople to Adabazar."

"What's your name?"

"Stepan Miskjian."

"Oh! Who doesn't know *emanetji* Stepan? I've seen you in the streets."

Stepan didn't recognize the man's voice or name, but he felt an instant kinship with him. As he shared his ordeal, he intentionally kept his back to the bushes, his face pointed ahead. He must have resembled a madman in rapt conversation with himself, but he wanted to protect the other man, his brother from Adabazar. "A month ago they took a census; they recorded where we come from, how many we are," Stepan continued. "The official in charge said that they were taking this information because it had been decided to return us to our villages — considering the deportation sufficient."

"Dear compatriot, don't believe everything they tell you," the brother snapped. "They're empty promises. Dogs are going to devour us. No trace of Armenians will remain. They took a census here, too, and supposedly sent the information to Constantinople. Not much time passed, and we heard that the *mutasarrif*, the governor, of Deir Zor had received instructions from the authorities to destroy the Armenians living under his authority, without mercy, and without exception. The previous *mutasarrif* had been a good man. 'I can't do that kind of work. I'm prepared to lose my position' had been his answer to the authorities."

The loathsome replacement was named Zeki Bey, the man explained. First thing, he traveled to the northern village of Shedadiye to convey the government's orders to the village elders. Then he hired two hundred locals and armed them with weapons and horses. The killing squad was told to wait in Suwar, a hamlet on a small river.

"The Armenians from the *amele taburu*, labor battalion, who numbered several thousand people, were put on the road at night, to where and why nobody knew," the man elaborated. "The first caravan that departed at night was completely massacred. Profiting from the darkness, a group of us escaped. From that day on, we've lived like this, hiding in the bushes."

Stepan slumped, too stunned to say a word. *Aman,* my God, he thought, *an entire caravan massacred?* During his exile, he'd watched guards beat his friends to death. Witnessed women raped, tents set afire, the screams of the sick inside. He'd seen atrocities, but not on this scale — this was the slaughtering of thousands at once.

Stepan checked the road for oncoming traffic. Still empty, he confirmed, so the man in the bushes resumed his story, how he had first encountered the butchers in Maraat, a tiny hamlet not far from the village of Suwar. "The *mutasarrif* searches out the remaining Armenians, gathers them into caravans, and sends them away to unknown destinations to be slaughtered," he said, before giving his final warning. "Tomorrow or the next day will be the turn of those gathered on the other side of the bridge."

Stepan knew that meant him. Nervous, he scanned the road again. In the distance marched two gendarmes leading a long line of captive Armenian females, just marshaled from their new Arab husbands.

"No more talking!" Stepan insisted, twisting toward the bushes and seeing the Adabazartsi for the first and last time. "There are gendarmes coming. Hide yourselves well. May God be with you. With all my heart I wish that you survive so that you can tell the whole of humanity what you know and have seen. *Yertak parov,* farewell."

Stepan turned back around just in time. The group was now in front of him.

"What are you doing here?" one guard demanded.

"My cart wheel was broken and I took it to town to have it repaired. I'm now on my way back to where I came from. I just stopped here to rest a bit." He then displayed his permission slip.

"Fine," the guard replied. "Get going."

He joined the group, all of them crossing together to the other bank. In a state of fear, he waited for morning, the time earmarked for slaughter. When it came and went, he struggled through the next day, and the next, until the days totaled a month. Stepan could barely sleep as he replayed the hidden man's ominous words. The rest of the time he spent circling the tents, scouring the trash for something remotely edible. Ever since his boss's customers had disappeared he had been on his own again, every so often unearthing an apple core or a piece of bread, rotten, but a treasure nonetheless.

Then one day, with the swiftness of a summer storm, the man's prediction came true. The guards were coming for them. "Dismantle the tents and push forward!" they yelled. Deir Zor camp was being cleared; the liquidation was in its final phase.

In the summer of 1916, as Stepan headed into the dreaded unknown, telegrams continued to tap into Germany's highest offices in Constantinople. "The persecution of the Armenians in the eastern provinces has reached its final stage," wrote Paul Wolff Metternich, the German ambassador, on July 10, 1916. "It is now about to dissolve

and disperse the last groups of Armenians who have survived the deportations."

It was not the first communiqué Wolff Metternich had sent to the
chancellor of Germany, Theobald von Bethmann-Hollweg, regarding
the Armenians. His stack of reports joined legions more from the con-
suls of Aleppo, Constantinople, and Damascus. Speculation in Aleppo
grew about whether Germany knew. No, some locals argued. Yes, others insisted, but they didn't act because of the alliance. Stepan was less
charitable, blaming both the Germans and the Turks almost equally.
"Not a leaf would turn in the empire without the Germans," he told
himself. Still, the Armenian patriarch of Constantinople believed the
German conscience wouldn't allow this to happen. "Mister Ambassador, I address myself a second time to Your Excellency, in the name of
humanity because I have the firm faith that a power as civilized as Germany can never consent to such a crime," he wrote. The patriarch dispatched this letter some two decades before Germany deported and
then slaughtered its own minorities.

At that time, Deir Zor was gaining in infamy. Through the testimonies of missionaries, counsels, army personnel, and survivors, a picture was emerging of a massive massacre in that region. In batches of
two to five thousand prisoners, the camp had been emptied, the Armenians led toward the Khabur River. In the middle of July, a large
caravan vanished into the copper-hued horizon. On the worn heels of
this, gendarmes arrested the remaining priests and leaders. In a telegram, Walter Rössler, the German consul in Aleppo, wrote that these
deportations were being carried out under "the pretext" of the Armenians being "dangerous to the health of others." On July 22, according
to more statements, yet another caravan left Deir Zor.

Stepan probably arrived during these July expulsions. In his mind he
always divided the massacres into just two: the one that preceded him,
as recounted by the man in the bush, and the one that included him.
Though foreign governments inquired into the state of the Armenians,
the official Ottoman response seemed to have quelled concern. A telegram from the American minister in Copenhagen relayed the Turkish position, "The Minister admits there is some suffering in Syria and

Lebanon as is also the case in Constantinople or Smyrna but denies that anyone has died of starvation."

It's a human flood, Stepan thought as the thousands inched along the Euphrates. Clouds of dust settled on them, one after another, coats of earthen paint on their olive skin, on their matted beards or braids. Stepan's convoy was being driven into a more solitary part of the desert, all of them stooped by the weight of their belongings, bent like half-moons above the dirt.

He'd believed Deir Zor was the end of the world. Now, though, he could see it wasn't. As he pushed farther east, it was like he was falling off the map into uncharted territory. There were no large towns within view, only a landscape with little or no water beyond the footpath. So far, he hadn't seen any oncoming caravans; the Armenians coursed in only one direction, southeast, like the river they followed.

Stepan looked out at the confusion around him, at the wizened bodies, barely balancing their bags, faltering on bloodied bare feet. He was in grave shape, but there were others who suffered more. Jerking his head to the right, then left, he looked for those especially in need. Then he helped carry their heavy sacks without expectation of anything in return. Many offered him food from their meager stashes anyway. He seemed pitiful, he knew. Over these past two years, he had become unrecognizable. He knew this without looking at his reflection, merely by reading the shocked expression of friends' faces when they saw him again and hearing the question that usually ensued: *Tun es Stepan?* Stepan, is that you? *Ayo,* yes.

Hours later, a hamlet came into view. Maraat. He could discern a few stone-and-stucco structures and not much else. It was August of 1916, two years since the draft had interrupted his morning meeting, almost one since he'd last seen his brother through the bars of his train. The escort had stopped so the guards could switch; their replacements menaced, with government-issued rifles draped over their shoulders, their heads topped with a *kalpak,* fur hat. They were Chechens, he learned, who had emigrated from Russia decades earlier. Fear engulfed Stepan as he remembered the words of his Adabazartsi brother in the bush, "These were the butchers. We were to wait here in Maraat [while] those who had gone before were 'cleaned up.'"

Stepan's caravan funneled into a large field pebbled with the footprints of another caravan. Recently vacated, he noted, the people already forced onward, the people already *cleaned up.* The guards busily designated a representative for every town. Onto large sheets of paper, these forty *mukhtars* jotted down their countrymen's names. More intermittent record-keeping by the administration, just as in Sabkha, two months earlier.

Before the *mukhtar,* Stepan spoke up and received his ration of flour — exactly one teacup. He stared at his hands, elated at that day's thimbleful. Nearby, an official was following the spectacle from his landau carriage. Through black spectacles framing deep-set eyes, the man observed the deportees, jubilant over the paltry handout. The twenty-eight-year-old official was indeed Zeki Bey, he learned, the one mentioned by the man in the bush. He tarried as the carts arrived, full of flour, the third time Stepan had received food from the government in a year. The *emanetji* didn't know what to think of the new governor, who was allowing them food when he had supposedly slated them for death.

The aid was from America, it was said. The plight of the Armenians had become something of a cause célèbre there; pleas for donations to the "starving Armenians" had appeared in the news. At the time, the United States remained neutral, and President Wilson's reelection campaign was in full swing, advertising his great success so far: "He kept us out of the war."

Soon, Stepan received more rations. Next came money. To collect it, he had to stand in line for two days in the burning sun and withstand frequent beatings from the guards. This was summer, when the sky remained clear, without even a cloud to soften it. The amount wasn't much — equivalent to two ghurush per person — but it was enough to raise his hopes some more. Two weeks after that came promises of more money. On that day, the town crier barreled into the sprawl of tents, his arms loaded with mail. Before the disheveled mass, he thundered out a message concerning the Armenians who had already "departed":

a number of them had received money at the post office, and any close relative of theirs could claim the funds.

The people became jumpy with excitement, heightened when another man emerged with a giant sheet of paper. He read off a list of exact amounts and names, enunciating a cacophony of *-ians*, the Armenian "son of"s that cap patronymics. Stepan strained to hear, everyone did, but he was wary. What had happened to the money's owners? No one was explaining more. Stepan concluded the owners were dead. More names rang out, and several people slunk forward. A line was forming. "Hajï Garabed Tarikian," the crier shouted now. "Ten Turkish pounds." Stepan couldn't believe it. That was Stepan's first cousin, the son of his wealthy aunt Hajï Abla. He had seen her in Qatma the previous winter, riding a camel with her other son. What had happened to Garabed? "I am a close relative," Stepan proclaimed and stood behind the others. The money could keep him afloat for days, if not weeks. When ready, the whole lot was led to a nearby governmental building, their cases settled by a tribunal. Inside, Stepan faced off with Zeki Bey again. For a man with such a grand reputation, he was short in stature. Protected by a ring of five, he held their destinies at the tip of his tongue.

"How are you related to Hajï Garabed Tarikian?" one asked.

"He is my maternal aunt's son," he said. "And I can confirm it by the testimony of our townspeople."

The panel exchanged glances. "You're too distant a relative," they proclaimed.

Nearly everyone's relationship turned out to be "too distant." Despite the letdown, the whole camp's spirits rose again with details of a confidential conversation between Zeki Bey and an Armenian he'd known back home in Everek, the town he'd lived in before the war.

"*Akhbar,* there's no need for you to tire yourself," he'd said. "You'll be together [with your family] in two months. The government plans to have all of you returned to your homes by the same route." The exhilarating news spread fast since it seemed credible. These were old friends from the old country, with a relationship predating this. Boundless joy passed from one deportee to the other. Stepan, too, reveled in the latest turn of events: the flour, the money, the promise from Zeki Bey, all road signs pointed to home. He traded his suspicion for blind optimism. He had to believe this. The alternative was just too devastating. *I will see my family . . . my mother . . . my sisters . . . my brother . . . my nieces . . . my nephew.*

A few days later, another summons rose above the din, calling the Armenian priests to a building. Twenty-seven of the frocked left, succeeded by their families, all moving past the settlement in two heaps. The remaining crowd observed, with envy. The lucky ones, they believed. Perhaps Stepan would be selected next. That night two priests from the first *sevkiyat* snuck into the camp, ashen with shock. They had nowhere else to go, and they recounted their horror: Their entire caravan had been massacred. A terror seized the listeners' faces and hearts; the nightmare hadn't ended. It had just begun.

By the next morning, the Chechen guards forced the caravan back onto the footpath. Leaving Maraat, the deportees changed course, away from the Euphrates — a worrisome turn.

The desert sky can look somehow artificial at dusk, like a painting. Reds and yellows and brilliant orange hues coalesce across an endless canvas. Stepan's route curved against it, his throat coated in thirst. Two hours on foot, six hours, ten, all without water. The brown terrain stretching outward like scars, the wounds of water deprivation. There was another way to the next stop that traced the river, but he wasn't being taken that way.

After a dozen or so hours northward, the cracked surface finally gave way to a splash of blue. A narrow river, Stepan could see, wriggled ahead, the Khabur. At one time, this minor tributary delineated the edge of the Roman Empire. He dipped cupped hands into the cool water, watched as it eddied around his fingers, and continued south on its journey to meet the Euphrates. Three days of scorching weather and marching, and the larger settlement of Suwar appeared, with its own military station. It was still August. Only thirty-three miles from Deir Zor, innumerable more from home.

Into one of the few buildings, the Chechens led the convoy. Previously a *khan*, now Chechen headquarters. Eight Armenian families were already rooming there. He saw a familiar face from home, a butcher also named Garabed, and his kin. The two exchanged information. Garabed told him that the massive caravan that had left Deir Zor right ahead of Stepan's had just passed through. Why Garabed hadn't been swept up in it, only God knew.

Stepan looked north. He saw nothing but level earth, flat as a pita. All around the region, the grass had been reduced to golden-brown shrivels, the impact of another summer drawing to a close. Arriving here, his old benefactor became despondent; he had burned his cart in a bonfire and slaughtered his own animals. He carved them up into portions, then passed the meat around to his friends like a last supper.

Undeterred, Stepan found eight other donkeys to shepherd. He was earning a mouthful of bread per day, the greatest payment. With grass so sparse, he had to herd them some distance so the long-eared animals, dark noses pointed downward, could sniff for food. Days of this, and then a group of men appeared near his spot, armed with swords, hammers, rifles, and cudgels. They edged closer, making him nervous, but they kept going. Within moments, another line followed, this one of boys, five to ten years old, wearing orphan uniforms, escorted by two mounted gendarmes. Stepan tried to count the little heads. One hundred, another hundred, another . . . one thousand children, he estimated. *Are those cudgels for the kids?* A few days later, his suspicions were confirmed when he overheard the guards boasting about their new game: how to kill children with the least amount of effort. By tying the little ones together, three or four at a time, they said, the sword could slash them efficiently with one sweeping stroke.

As the rumors about what was transpiring upriver grew, Stepan thought he couldn't bear much more. This time, the stories came from a boy whose caravan was wiped out. He'd survived, he explained, by hiding in a pile of stolen goods and then crossing the river. The massacre had happened behind a hill, where Zeki Bey had unfurled a large carpet and demanded the people's jewels and gold coins. Then he had made them strip and shot them all. The more fortunate ones, like the boy, began to stagger out and went on to Aleppo, informing foreign dignitaries of the awful truth. With these testimonies, the American consul there wrote that the extermination of Armenians was near complete. "Aside from less than a hundred that escaped, and about 250 small children that were left running in the streets of Deir like dogs, the entire 60,000 Armenians were wiped out within a week," J. B. Jackson wrote. This occurred immediately preceding Stepan's arrival. This seemed to be the caravan that Stepan's had been trailing, the one whose campfire rings he'd found at each stop.

At summer's end, Stepan set off to find more pasture, which proved harder than ever. Leading his donkeys, he passed the spot of the doomed orphans and kept going. No food seemed to remain anymore for man or beast. At last, near some rocks, he stumbled on an isolated clearing with a bit of brush and stopped. As the donkeys hungrily grazed, he had the distinct feeling they weren't alone.

He was right, and now he saw a naked man motioning him over. The stranger was in grotesque shape. His head had been split open, his arm and back gashed. The man struggled to talk. "I am from Marash and my name is Panos." He'd been in the Suwar camp, same as Stepan, this last gateway before the desert's end.

"They got us on the road again," Panos related. "After traveling for an hour we had to stop. They separated all the animals. Then they separated us, the men from the women. New clothes are put to one side. They then fashioned a kind of rope by tearing the old clothes and tying the victims with it. After walking for about ten minutes, they delivered us behind a hill where there were several hundred savage and hideous people waiting. They began first by taking whatever clothes we still had and then killed ruthlessly, with their special instruments: Arab swords, clubs, and hammers. I was bound in a group of three."

Panos paused to catch his breath. Stepan must have lost his, too, as he tried to picture this sequence of violence.

"I was hit on my head by a club," he continued. "Before falling, the scene was horrifying. We were too many. People fell on top of each other. I saw that I was lying in blood. Almost all were dead. I heard an outcry from afar. Two other people had fallen on me. Slowly, I raised my head a little. There was no one left around us. I could hear the shouts and cries, the death agonies, the heartrending laments of the victims. It was a horrifying scene. Whenever a voice was heard the assassins ran to smother it with the blow of a sword. Hell! Hell! Minutes of overwhelming suffering, which quickly ended: And then all was quiet. The desert had turned into a cemetery, a necropolis."

The story mirrored the one from Deir Zor Stepan had heard in the brush, about the male labor battalions. But this wrenched Stepan's heart even more. It wasn't just a tale; the man was bleeding before his eyes. Stepan thought of the ones already dead and his friends camped in Suwar, to say nothing of himself. The disfigured man struggled onward, "Luckily my wounds weren't deep. It was amazing that in the midst of my terror and fear of dying, I had forgotten my pain. I was able to untie my bindings, free myself from the entanglement of my friends' bodies, and raise my head a bit, and look around — what a horrifying sight! Some were dismembered, beheaded, others without arms, or disemboweled."

Peeking between his dead friends, he'd watched the executioners group the women into tens and then pick up their weapons. The silence was broken by high-pitched squeals: "*Mayr! Mayr!* Mother! Mother!" For the next two hours, the women begged for help until their cries, too, disappeared. Among the dead, the man lay frozen until nightfall, when he slid out from his crypt and tiptoed to this spot. He needed clothes, and he asked Stepan to get him some.

"With great love," Stepan replied.

Returning to the encampment to collect the garments, he reflected on the dire harbingers he'd already witnessed. Several times, while he was drinking water at the river's edge, corpses had floated past him downstream. Tied in threes and fours, bare like wayward logs, ropes entwining one to the other. Stepan didn't want to believe, but listening to this man's account, he could no longer ignore the signs: death did wait on the road up ahead.

Welcome to Syria

2007

ON THIS BARREN STRETCH between Turkey and Syria, I felt a kick of stomach pain. I doubled over, dropping my head between my knees; the nausea intensified. Then came another wave. Still stuck in international limbo, I felt like screaming. I never foresaw that this trip across the border into Syria could go so wrong, that I would end up hitching a ride with a carload of strange men, that I would be pulled over, effectively imprisoned between the two countries and brown mountains.

The heat was stifling here, the sun's lashings excoriating. My skin was reddening into blotches, hypersensitive because of an antibiotic I'd taken the previous night. Nearby stood the car's other passengers. Shifting their feet, they, too, seemed impatient; our driver had been in that nearby building for what felt like ages. *Are they detaining him? What has he done?* How I longed to be singing my way across eastern Anatolia again with Jemal and Baykar. I was half delirious, half paranoid, certain only that I had to extract myself from this situation. On wobbly legs, I crossed the hot asphalt and strode through the ripples of heat to the Turkish sentry stationed outside the office.

"Do you speak English?"

"A little," he said.

"I'm not with these men. I just met them. Can I continue on by foot?"

"It is still very far," he said.

Defeated, I sat back down in the blistering sun. A lone branch of shade barely covered an arm, and I contorted in vain into its silhouette. Finally, our driver came barreling out the door, rushing toward us. He motioned for us to get back into the car. After we'd piled in again, the station wagon sped the remaining distance, the road winding at a slight grade. I couldn't believe that I'd thought I could walk this in my depleted state. Up ahead loomed the Syrian checkpoint, spanning the road like a bridge. *The last hurdle,* I thought.

In that moment, another officer stepped into our lane and waved us down. Beside the car, he addressed the men in Arabic. At once, my fellow passengers retrieved their passports and gave them to the officer, signaling for me to do the same; at least, that's what I assumed their gestures meant. Leaning forward, I handed mine to the official. After a brief inspection, he began to return the precious documents, fanning them out around the car like sold-out concert tickets. Nearly incoherent from my sickness, I extended my hand, expectant. Instead, he directed me out of the car with a stern voice. Now I really started to panic.

Through the windows, the car's men stared at me as I extracted my heavy roller bag from the trunk. They were all strangers to me, but they were more comforting than this guard, who now led me to a spartan governmental office. There, an official sat behind a desk busily questioning another foreigner, a young girl in her twenties. When finished, he turned to me, his demeanor pleasant as I took the hot seat. I tried to appear calm.

"Why have you come to Syria?"

"I want to visit the country."

I kept quiet about retracing the deportation route. An Armenian filmmaker had shared with me his experience of being granted permission to shoot a documentary about the genocide and then having it revoked. I couldn't jeopardize this trip; I needed to find the family of the Arab sheikh who'd saved my grandfather. I didn't even know if this was possible, but I had to try. So I had decided to enter as a regular tourist instead of a journalist. In essence, that was what I was. I just had different landmarks on my itinerary than most sightseers.

"Why Syria?" the man asked.

"I am also visiting someone." Before I left, a family friend had insisted I stay with her aunt, a resident of Aleppo, so I wouldn't be completely on my own. I argued with her for the longest time, maintaining that I had plenty of experience traveling in foreign countries and handling unexpected situations. "It will be easier if I just check in to a hotel," I said, not wanting to impose, but she wouldn't hear of it. As I sat in this office now, trapped, I felt immensely grateful for her persistence. *Someone is expecting me,* I told myself, *someone will know if I go missing.* The official flipped through the pages in my passport and lingered on one, covered in stamps. "Why did you visit Brazil?"

"For vacation," I said. I found this questioning so curious: How could my trip to the Amazon, where I fished for piranhas, now be suspicious? I was no fearless femme fatale agent, though I could understand the mistaken identity. In Brazil, I ran at the speed of light from a tiny spider and courageously posed next to a restrained baby crocodile. The inspector's mistrust was astonishing, yet real, and he pressed me on the specifics for the next hour, almost down to the number of bugs that had stung each leg. I was trying to appear casual, but my arm began to shake, so I braced it in between my knees. Finally, a middle-aged man with cropped hair appeared at the doorway and spoke in Arabic. I recognized his strong voice and lilting cadence. It was my Syrian translator, Levon (whose name I've changed). I'd never met him in person, but I had spoken to him several times before my arrival and, thanks to the urging from my family friend, had requested he meet me on the Syrian side.

"Parev," we said to each other in Armenian. "One minute," he said, and he vouched for me to the official. Into his hands, I was soon released, and we drove the remaining miles to Aleppo, passing arid land dotted with square homes. I'd learn later that this suspicious treatment at the border was standard fare for foreigners entering the country. But what had begun as routine escalated into something else the farther I traveled into the country. In the car, Levon quickly shifted gears from rescuer to tour guide and began to tell me about their thriving community of a hundred thousand, the majority living in Aleppo. Many

Syrian-Armenians were originally from Anatolia, Levon explained, and had settled there following the Armenian genocide.

After we wound through the sprawl of the city, Levon dropped me off at my contact's spacious house, where I crashed for days, vomiting, curled up like a worm. Her poof of a white poodle kept tiptoeing into my room, curious about the visitor, her nails against the tile always foretelling her arrival. I had a permit to remain in the country for only two short weeks, and precious days were passing as I slept, febrile, the air oppressively hot. I couldn't imagine lying outside in this state, exposed to the sun or snowdrifts, like my grandfather. Despite a soft bed and a gracious host as nurse, I was miserable. At last, that Monday, my powerful antibiotic finally chased down my temperature. Not losing another minute, I climbed into the back seat of a taxi, an Armenian driver behind the steering wheel and Levon in the passenger seat. Together, we set out east to visit the region that was once riddled with camps, the places where my grandfather had lived and nearly died.

The Desert's End

1916

"YOU ARE BEING SENT BACK HOME," a Chechen guard announced. "The deportations have ended, but to avoid overcrowding on the roads, you will return in small groups."

By now, the tens of thousands of exiled Armenians knew better. They didn't cheer, not like before; instead, a paralyzing fear clotted their throats. The wait in this Mesopotamian camp of Suwar finally seemed over; August had become September, and September had become the end. They knew what came next. At the dirt podium, the official called out specific villages and towns, the first supposed convoy home. Stepan listened for his, Adabazar. The memory of living comfortably with his mother and siblings in the days before deportation and war scarcely felt real anymore.

The cracked earth surrounded them, with few signs of life. This was the desert's end, for him and everyone else. Beside him, thousands stepped forward on bloodied and bare feet and assembled into a line. Stepan couldn't believe their passivity; no one was resisting, not one person, the fight gone. In their arms, some carried bedding, in hopes of needing it, in hopes of a future. He studied their faces, noticed their quivering lips. No words were coming out; the procession formed in absolute silence. *Are they whispering their last prayers to God, or are they delirious?*

At last, the trembling convoy waded out to the opposite bank of the narrow Khabur River, beside armed guards. Then the official named another town, and the scene repeated itself. Some men around Stepan couldn't take it. Alone like him, and starving, they declared, "Whenever or wherever we're sent, we're going to die. We're condemned to die. It would be better to die sooner than later."

"Let's die and be saved," said another.

Sometimes, Stepan thought that too. But he wasn't ready yet. He wanted to live, even for one more day.

Another three weeks elapsed as Stepan waited for Adabazar to be called. September turned to October, and the guards barricaded in the Armenians left behind. No one could roam beyond the tents clustered outside the hamlet. Not even the animals, which were now tied up to starve. *Asdvadz! Asdvadz!* Everyone left began to cry out, summoning God. Some retreated into their tents to wait. On the floors, Bibles were butterflied open, pages turned. The pious bent knees and clasped hands, locked in perpetual prayer, their favorite psalms read aloud. *Though I walk through the valley of the shadow of death, I will fear no evil: for Thou art with me* . . . Some, already starving, stopped eating, remembering how Christ had done so and how they used to purify their souls during Lent. The abstinence gifted them with something unexpected — control, when there was none.

Names of new hometowns thundered down onto the crowd, and thousands more people stepped forward. Stepan felt cornered. Still no Adabazar, so he stayed put. Some Armenians tried to escape. He watched the Chechens gun them down. Others attempted to buy their way out with a *baksheesh*, like Khachig from Bursa. Distinguished, with a long mustache, he'd been the prestigious director of his town's Singer Company, the sewing-machine manufacturer. Now he limped over on his lame leg to meet with the head of the Chechens. "I will pay eight hundred lira gold pieces to be spared," he pleaded. The *müdür* accepted these terms — and then had both Khachig and his sister-in-law killed.

Once again, the guards focused on the most affluent deportees. One by one, their names shot through the air, including that of Antranig

Giumiushian, Stepan's friend from home. Since the village of Sabkha, Stepan had been on the road with him. Antranig now joined the other well-off exiles, numbering six hundred. The chief Chechen had a deal to offer on behalf of the district's governor, "One thousand Ottoman gold pounds for all of your freedom." Back in Adabazar, Antranig had been a highly respected textile manufacturer and had been voted onto a prominent education committee. As the patriarch of the family, he had a lot to lose; his wife and two daughters accompanied him, along with his son-in-law, Garbis.

After scraping together the requested gold pounds, Antranig and the other notables handed it to the Chechen. Two days later, the mediator relayed a message from the governor, Zeki Bey, "The man wants two thousand. Yet it is your decision what to do here. If you want, you can collect another thousand gold or you can take back what you gave. See, here is your money." He fanned out the gold pounds as if he were ready to return them. Even from afar, Stepan understood the disingenuousness of this gesture. Hopeful, Antranig and the others emptied their pockets again, only to be imprisoned the next day.

"Ahead toward the desert," a guard yelled. That meant everyone. Stepan struggled to his feet as the remaining six thousand or so stirred into formation. All the others were gone. This time, resigned, many left their tents standing. Still others clung to their possessions, like Yeghisapet Sarian. Stepan knew her well, knew how much she cherished her five bags. For days, in exchange for lodging, he had been helping the wealthy woman, assisting with her woven Oriental carpets and fine clothes that had somehow been carted from her home near Adabazar. The convoy had begun to move. Stepan was hurrying to hoist her *bohcha,* as heavy as the dead, onto the donkey when a Chechen guard approached.

"Let me give you friendly advice," he said to Yeghisapet. "Don't take this load with you. The roads are poor and you'll surely be troubled. It would be better if you left them all here with someone you can trust, who will then send them on after you."

Stepan kept his eyes on Yeghisapet. Her expression was wavering.

He knew the guard was lying, that she wouldn't need anyone to send her the rugs. Yet she began to motion to a tall Chechen with a beard. His name was Shakir. Before the war, he had been an employee of a company that regulated the empire's tobacco, and he was married to an Armenian. Yeghisapet thought him honorable and asked him to assist.

"With the greatest pleasure, madame," said the man in a sweet, reassuring tone. "When you reach your destination, write to me and I'll make sure all is sent to you."

After Yeghisapet handed him her life's possessions, she and Stepan joined the caravan, leaving Suwar at their backs. Along the soggy bank of the Khabur, the fur-capped sentries led the procession, periodically reminding them that they were headed home. One of them singled Stepan out and told him to take care of their donkeys. Not daring to refuse, he scooted behind eight of the animals that swayed with luggage piled high, loot plundered from the deceased. He and six other Armenians prodded them forward. The convoy lumbered along, passing truly gruesome scenes. A friend from home, Harutiun Atanasian, saw them too. In fact, the publisher couldn't move without coming upon another corpse. The dead were scattered on the road like stepping-stones to the massacre site that now appeared, the bodies akimbo, soaked in gore. The convoy kept going, the sky turning to mud. Stepan had no idea where they were, only that they were halting for the night.

In the pitch-black, he could barely see in front of him, yet he managed to erect Yeghisapet's tent. Then he thanked God for his surviving another day. In an instant, a great flare illuminated the desert. *It looks like a sword,* he thought. For a few minutes, he scrutinized the image. Then, with another flash, it vanished as quickly as it had come. What was that? They all marveled, each one reading the sky the way an astrologer interprets a chart: "We are going to be slaughtered!" a few said. "The sword means we are to be massacred." *Voch,* that wasn't true, argued others; there would be no massacre because the sword had disappeared.

Eventually, the clarity of morning came. The sword meant nothing, no rescue party, no army galloping in to save them. Stepan and the others roused to more of the same march, tracing the spirals of the tributary farther north. It was only Thursday of the longest week of his life.

Up ahead, a mound rose up about one hundred feet. Piles of small black
boulders rocked on the top. It was the lone rise for as far as the eye
could see, a solitary ship in a vast ocean. Is this what the wounded man
had spoken of? Had warned him about? His heart quickened. The folks
in front of Stepan were tapering to a stop.

They were somewhere between Suwar and the village of Shedadiye.
The streaming of water rushed nearby, sharpening and carving the bank
by sheer torturous repetition. The swooshing was loud — or it seemed
that way to those who stood on its shore and contemplated their last
moments of life. Stepan and his companions settled beside this high
mound. Just east of them, along the river's curved spine, squatted the
tents of three hundred guards.

Zeki Bey had visited this area earlier and reminded the guards of
their duty by using a prop. "I learned that you on occasion take pity or
bribes and save people," he had lectured, lifting up a two-year-old boy.
"Even this innocent should be killed without mercy. A day will come
into effect when they come, seek those responsible for these killings of
Armenians and will draw revenge." Then he tossed the child high up
into the air several times before bashing him against the ground.

These guards had already killed and were about to kill again. Of the
more than two hundred thousand Armenians forced farther east along
the Euphrates to the Deir Zor area, only this six thousand remained.
This is what was left of Stepan's caravan.

Two more days of wasting away and waiting. Now the steady skyline
was interrupted by movement. Dozens of children were walking toward the encampment, still dressed in their uniforms from their orphanage, the scene an echo of the earlier group. Stepan scanned the
line; two hundred long, it seemed, with matronly women marching beside them. Must be their caretakers, he decided. Which orphanage did
they come from? Aleppo, he guessed. Harutiun thought Deir Zor. That
Saturday, the orphans' caravan came to a standstill, and they pitched
their tents thirty yards away. Their overseers had ordered them to stay
put, but somehow, two courageous ones slipped into the somber night
and recounted their ordeal to Stepan and the other adults. The chil-

dren had left Deir Zor ten days earlier and been starved for the past three. There had been more. Many more.

In the morning, the guards forced the little ones onward. Stepan and Harutiun watched, knowing they'd all be killed. Two hours later, the guards reappeared, as expected, the task completed. After the war, when it was all over, the police commissioner of Deir Zor would describe two different massacres of orphans. He had overseen them, he testified. One was a large-scale operation, the drowning of two thousand children in the Euphrates River.

Before Stepan's eyes, the next generation of Armenians was being extinguished. His dreams of having his own children too. Another deportee who had marched on this same stretch told of another almost inconceivable sight, farther north. "We walked. We passed the half of the way, when we arrived at a tomb," recounted Aram Bouloutian, who was around twenty-four years old at the end of 1916. "We noticed smoke. On that path, we met Arab peasants. We called one of the Arabs, asked, 'What is that smoke? It smells too.' He replied, 'Don't ask. Don't ask. Those heartless people [took] twelve hundred Armenian boys, aged ten to twelve, poured gas on them and burned them. It has been burning for five days, and it is not ending.'"

In that part of the desert on that fall day, the guards waited, the stench of rotting corpses lingered, and Stepan's life was extended for a few more blessed hours.

Whatever happens will happen, he thought. *I'm not afraid of death.*

At the bottom of his sadness, Stepan had an epiphany. It was so clear, he praised God right then and gave thanks. For a long time, Stepan hadn't been able to see the positive side to anything. Yet he had had a revelation: he would be the sole person killed in his family. Only him. Somehow, according to his latest information, his mother and four sisters and brother had remained in a camp in Anatolia and avoided the death march to the desert. That meant they had a chance, and he'd most likely be the only victim. At least, that's what he hoped. *I'm lucky,* he consoled himself.

This moment, the nadir of his life, also had significant personal meaning. It was the anniversary of the death of his father, Hovhannes: September 25, 1916, in the old Julian calendar, October 8 in the Gregorian one. He had died twenty years earlier, when Stepan was only ten. All the time, Stepan had missed him, but not for much longer. On this Sunday afternoon, the grown man closed his eyes, pressed his long thin hands together, and prayed, *Dear Father. Today I will be with you.*

The time had come to line up. Through his fog of thoughts, he could hear the guards scream: "Everybody, ahead!" The Armenians rose to their feet, obediently standing in front, behind, and beside one other. The last caravan had formed. *Why should I leave my twenty ghurush to these butchers?* he asked himself.

Taking out his precious change, Stepan spotted an Arab merchant, one of many who had traveled to this area with sacks of dried grapes and dates to be sold at inflated prices. Stepan approached him and bought three hundred dirhams worth of raisins at about four times what they would usually cost. Stepan savored them. Skeletal and nearly naked, he dropped the shriveled sweetness of each one into his mouth. He could feel the raisin plump with his saliva, almost come to life, and taste the pop of a sugar rush as he bit down. Then he positioned himself among his friends, all his money gone. Soon, they would see the sign.

Ten minutes passed like this. The quiet was stifling; it was still enough to hear his own breath, his own beautiful breath. Finally, the guards motioned. Another raisin plopped into Stepan's mouth, and his legs, thin as skewers, began to move. Where? He didn't know the exact coordinates, but this time, he knew his final destination.

I have one or two hours of life left. One or two . . . Almost possessed, the six thousand remaining deportees ambled ahead and did as they were told. One foot in front of the other, with no one bucking the tide. Another fifteen minutes elapsed. Now an order to stop. Others near Stepan spoke aloud their thoughts, "Yes, they've started, they're going to start . . ." The damned stepped forward, some numb, others screaming and crying, unhinged. Every minute was their last. The old, the young, whole families linked together, threading arms around other arms, saying goodbye to one another, yelling, *Asdvadz! Asdvadz! Asdvadz!*

"Calm down, quiet, calm down," the tormentors yelled.

Silence fell. As if looking at a map, they read the names of the densely populated Armenian towns in eastern Anatolia. "Everyone from Adana, Hajin, Marash, Elbistan, come this way!" It was a piercing sound. So far, no Adabazar.

Nestled together, the captives were torn. Murmurs drifted from one to the next: Should we go? Or stay? A line took shape. Some rolled the dice and went. Other residents stayed behind. *Death is on both sides,* Stepan thought of the caravan, now divided into two groups of three thousand. Stepan remained.

"You will go tomorrow," the guards said to those who stayed behind.

"*Eh,*" a few exiles said, "we gained one more day."

Stepan's eyes followed the damned as they shuffled across the dirt, their silhouettes disappearing behind a hill. An hour creaked by. Another hour. As evening fell, some women from the previous group returned and pitched their tents a ways away. One woman managed to sneak away to Stepan's group. Hysterical, she recounted what had transpired: "They lined up all our men and slaughtered them and moved us back here, saying that tomorrow they'd take us with you." Amid the screaming, crying, and flailing, the blueprint of the massacre was revealed: The men were separated and led to their death, the women targeted next. It was just a matter of time now. Some nine hundred miles from home, in the middle of the desert, in the middle of nowhere, Stepan reconciled with his present too: the last hope of survival was gone.

That night, no one wanted to sleep away his or her last hours. Above, a full moon illuminated the sky, casting some light for the several hundred families left. Stepan stood around, his senses sharp. In small clusters, people were starting to whisper. Close enough to eavesdrop, he learned that some were plotting an escape. Hearing this awoke a yearning inside. He wanted to live. Or, rather, he didn't want to die here — anywhere but here. The desire raged inside, desperately now. He sought out his close friends, Harutiun and his wife, Lusi, and shared the plan. Lusi became frenzied, imploring her husband to go without her since she feared she was too weak. "It is useless," Harutiun replied. "Wherever we go, whether we stay or run away, the end result is dying. I'll die here with my wife."

How about the other exiles, Stepan's friend Hajï Bedros Dimijian and his old neighbor Melkon Lorentsian, the one who shared his food? *Voch.* Next he reached out to another comrade. *Voch,* he could no longer walk. Stepan understood why many decided not to flee, but he felt sickened by their defeat and by having to leave them behind. A boundless desert swept in all directions, their exact location a mystery. Besides this hill, nothing but mounds of small stones dappled the ground, used by locals as primitive road signs but pointless to hide behind. Blackness blotted out the contours of sand that stretched to the horizon. At the very least, it was a weeklong journey to the safe haven of Raqqa, made more perilous without provisions or arms. Undaunted, Stepan overcame his usual hesitation to ask others for help and he whispered to the group huddled around him, "Give me a water bottle if you have any." His friend Roksanna answered, "Wait, Stepan Akhbar, we have a bottle," and she placed a bottle filled with two cups of water into his hands. He tucked it into his pocket, feeling grateful for the two cups. He was also sad, knowing what this meant for her. Then Yeghisapet Sarian, who had held on to her possessions for so long, reached into her pocket and bestowed a precious French gold coin. "I am going to die," she declared. "What's the use?"

She had given up, despite the rumor circulating that the women might be spared. Just in case, Stepan penned a final letter to his family. Maybe if some of the women did survive, one of them would pass it on. *Don't worry about me,* he wrote. *If I die, I am consoled by the idea that I am the only victim of the family, and for that I am glad. Goodbye. I kiss you all.* He copied this three times and distributed the notes to his friends.

"May God be with all of you," Stepan said.

"May God be with you," they replied.

Saying goodbye, he kissed the hands of the elderly and shook the hands of the young.

From afar, Harutiun and Lusi watched the first pack of escapees. In the darkness, their shapes were barely discernible, visible only to those who knew where to look. First, twelve men edged out, with small bags. Glancing back, they moved their hands from their foreheads to their chests to their shoulders, the sign of a cross, before vanishing behind the veil of night like phantoms. Minutes later, fifteen others followed. Soon, ten more.

It was late, perhaps eight o'clock. Stepan edged toward the outer part of the camp, pretending he had to urinate. "Do what you need to do right there," shouted the guards from the camp's outer ring. "Don't go any farther or you will be killed." He retreated.

Harutiun had been biding his time with his wife, Lusi, the one he loved so much, the one he wanted to stay with until his last moment. Lost in thought, he pondered whether the women would be spared. Not even rumors promised to save the men. Harutiun knew this first-hand, having learned about the Chechen's directive: to kill all males over age twelve. Finally, Lusi couldn't take it anymore. Turning to her husband, she said, "'Stepan is leaving, you also should leave ... If you think that by escaping you'll be saved, go!"

He didn't want to go.

"You get away and be free," she implored. "We'll find each other after."

The two began to sob and held on to each other. Lusi was right, Harutiun knew. It was a decision that he would revisit for the rest of his life. One made in just seconds. Now there was so much preparation. In what direction would he walk? To Raqqa, he decided, and he procured some water for the trek. The tears rolled down both their cheeks. He couldn't speak; neither one of them could. In silence, he separated from her into a blurred world.

It had been three and a half hours since Stepan's last attempt. The majority of the guards had now fallen asleep, their breath giving the only ordered rhythm to the night. He gestured that he needed to relieve himself once again and proceeded a few feet. No voices protested, so he advanced farther. He spotted the outer chain of guards ahead. They were sleeping too. To sneak through, he had to pass between two sentries. He lowered himself onto his hands and knees, inhaled deeply, and held his breath as he crawled past their feet. "Have a sweet sleep and a nice dream," Stepan said to those he had left behind. And then he began to run.

My Shadow

FROM ALL ANGLES, the sun was bearing down. Even with the car's air conditioner cranked high, I was still hot. And thirsty, too, as yet more desert passed outside, and occasionally small earthen homes. It had been a week since we'd left Aleppo, and nearly all of Syria lay in our rearview mirror, including those hamlets whose names I knew well — Qatma, Meskene, Dibsi, the onetime camps. Now east of Deir Zor city, we were speeding toward the Iraq border. This was the stretch I had been waiting for; this was the terminus of the deportation.

My translator, Levon, pointed toward some greenery off to the right. "That is the Khabur River," he said. "Quick, pull over!" I exclaimed. We were passing the waterway that led to my grandfather's massacre site. As our driver, Antranig, swerved hard onto the shoulder, I noticed a station wagon trailing us by a dozen feet. At the slam of Antranig's brakes, I flew forward in my seat; the station wagon also abruptly stopped, narrowly missing us. It pulled off the road too. My heart rate quickened. "Are we being followed?" I asked my translator, though the answer was obvious. "Yes," he said matter-of-factly. As the driver killed the engine, I peeked out the back window. Two or three plainclothesmen, it seemed. This must be the Mukhabarat, I told myself, the Syrian secret police; it had to be. They were parked behind us in this remote spot, their intentions unclear. I hadn't noticed their vehicle that morning when we left our hotel in nearby Deir Zor, now a noisy, congested city. Levon must have seen it, but he had chosen not to tell me, which I understood. After decades of Assad's rule, Syrians had grown somewhat inured to the surveillance, it seemed. A foreigner being followed was nothing special; it was almost standard operating procedure. However, resentment toward the regime was increasing, just under the surface. In a few years, it would explode.

"*Salam*," I said to the policeman as I exited the car. I was trying to act like any tourist at a panorama, camera in hand. At this spot, the placid Khabur met the Euphrates, the intersection marsh-like and swampy with long swaying reeds clumped like small islands. "Okay to take a photo?" I asked. Levon translated, and they indicated it was, so I snapped away. "*Shukran*," I said, thanking them.

We got back in the car, and they did too. This was beginning to feel like a game of monkey-see, monkey-do, and I briefly considered touching my ears but then thought better of it. We drove a few minutes, the officers behind us, and we all halted again in front of a house and a mound of dirt. It was one forty-five in the afternoon on a Friday. I didn't understand the purpose of this detour. "Why are we here?" I asked. "Follow me," Levon said mysteriously, traversing the road. I could feel the stares of the intelligence officers, now leaning against their car, as Levon led me to the mound. Shaped like a horseshoe, it looked like buried trash. "This is a massacre site," he said bluntly, without emotion. I stood there, frozen. "There are bones in there," he added, like a man who'd been here many times. He was scaling it now, digging his hands in as he climbed. The earth was not smooth, with ditches in some parts and piles in others. Weather-beaten, it seemed to have molded to the remains underneath.

The place was called al-Busayrah and was once on the deportation road. I tiptoed closer, careful to avoid any raised area, wanting to respect the dead. Nothing marked this place. No eternal flame or cross or placard. The hill of bones sat next to a main thoroughfare. How easily I could have kept driving, unaware of the thousands that had died there, forgotten in this human heap.

"Go ahead, take a picture," Levon said. I glanced over my shoulder; the Mukhabarat men were observing, wordlessly. I hesitated. I didn't want to do anything that could be construed as wrong and end up interrogated — or worse. Instead, I did something else. I clasped my hands together like my grandfather, closed my eyes, and whispered to those trapped forever in this desert's graveyard, "Rest in peace."

Tell the World

1916

FOR HOURS, Stepan sprinted across the desert floor. In the darkness, he couldn't always see the small shrubs, and the natural variations of the earth that stole his balance and twisted ankles. Yet a frenetic energy propelled him forward, deeper into the unknown. Finally feeling safer, he dropped to the ground. Inhaling the cool air, his breath labored, he considered what to do next.

With only two cups of water, he had deduced his chances of survival were slim. In every direction, there was sand and only sand. During his escape from the caravan earlier that night, he had just run. It didn't seem to matter where; he had to get away from that hill, place a wide distance between himself and the Chechen killers in the fur caps. Others had escaped in pairs; they could help each other find the Euphrates River. But he was alone.

"Bless you, God, that I'm still alive, all the way here," he said aloud. "But what am I to do after this?"

Stepan wondered if the butchers noticed there was less to slaughter. Were they hunting him down? His dear comrades . . . were they lying dead in a pile? Stepan rose to his feet and continued in what he hoped was a southwesterly direction toward the riverside town of Raqqa. A guess in the dark, that's all it was.

As the new Monday sky filled with yellows and blush tones, the warmth of October 9 steadily strengthened. Soon it would be sweltering. With just a few rags hanging from his bony chest, the *emanetji* stood bereft of protection from a sun that blistered and peeled skin like a parboiled tomato. He consulted the sky, observed the sun's position, nature's gift to lost men. Turning his back to the rising colors in the east, he could confirm he was headed in the right direction. Before fleeing on Sunday, he had asked some friends how they thought he might get to Raqqa, the ancient town on the Euphrates that he'd regretted leaving. Carefully, he had rendered a makeshift map on paper and sketched out a route that would take maybe eight days. Now Stepan was bone tired. *Must rest, and hide.* He mounted a small hill, stacked some stones in a circle, and lay down in the center. In his primitive fort, he shut his tired eyes.

Unbeknownst to Stepan, his friend Harutiun Atanasian wasn't far away. He had fled the hill massacres not long after Stepan. All night they had each been racing toward Raqqa, the journey proving too much already. Having forgotten his water, Harutiun fell down at dawn too, exhausted. Both men now struggled to survive in this remote tract of desert.

Stepan slept until the heat bore down so intensely, it slapped him awake. He looked around, surveying the golden landscape that seemed to have swallowed him whole. The sun sat brightly overhead, illuminating the futility of it all. Already, he had passed a handful of small rises. Now, on the precipice of entering a more intimidating stretch of the desert, he was at a complete loss. Then the solution came to him, the only thing, other than dying, he could do to escape: he would ask for help, the same way he had before. With his torso erect, he clasped his hands together, closed his sunken eyes, "Glory to you, my God. You saved me and brought me this far. Protect me from all danger. I have one French gold piece in my pocket and a small bottle of water. Save me before they are gone. Whatever happens, may Your will be done."

Elsewhere on the caravan trail, many Armenians had spoken similar prayers and then lost their faith when they were left unanswered. Stepan's mother, Hripsime, for one, couldn't comprehend why a benevolent God would allow such pain to exist. "Everything is false regarding religion," she would say later in life. Stepan's faith, by contrast, grew like a tree in this barren land, rooting deeper with each perilous moment, each time he lived through another day. Finishing his orison, Stepan reached for his water bottle and carefully rationed a drink.

Some relief. It was a miracle he was still alive, he knew.

Around this time, the caravan he had left was summoned. Specifically, all males over the age of twelve. At a spot on the Khabur River, the butchers shot their prey. Only a few boys escaped the firing line, saved by masquerading as girls. In the frenzy of gunfire, the women gaped as their loved ones were killed, an agony Stepan considered worse than death. The rumors of the women being spared turned out to be partially true. The guards killed only a few females outright, but the rest were left, naked and hungry, to die in the desert, Harutiun's wife, Lusi, among them. That Sunday was the last time he saw her, the last time he told her he loved her. The earth eventually claimed the dead, the feminine wrists and sweetheart lips, the Adam's apples and clavicles. Their fragments buried under a mound, their only testimony and voice to the crimes. The songs wept on the road foretelling their fate:

> I rotted and remained in the desert of Deir Zor.
> I remained and became a meal for the crows,
> Oh, Mother! Oh, Mother!
> Our condition was lamentable,
> At the time we were in the desert of Deir Zor.

Stepan ventured forward again. Before too long, he encountered a hillock. In cresting it, he could view his environs, the land so still it seemed to slumber. On his descent, though, he detected something in the distance: four people striding toward him. Quickly, the *emanetji* assessed the threat. These days, he made judgments based on intuition, the inner voice that guides. So far, his survival had been largely due to this.

Somehow he knew not to fear these men, knew that they were Armenians like him.

Stepan called to them and flailed his arms, and they called back and returned the gesture. Slightly diverting their routes, they all converged. The men hailed from Fernuz, they explained, near the Cilician town of Zeitun. They, too, had escaped the night before. Stepan felt relieved not to be alone anymore. Until moments later, when seven or eight men appeared nearby, dressed in the long coats typical of Arabs. Panicked, Stepan shot off at a run. The men from Fernuz followed. Observing this, the strangers sensed a misreading of the situation, and they shouted at the runners. More calling out ensued, and through the confusion, the men finally understood that they were all in the same situation.

Now the party of eleven moved forward. Stepan felt like dropping. Not far, he spotted some stones, largish in size, and he positioned them so they could all hide — or at least try. For Stepan, several days had already passed without any food, much longer since he'd had more than solitary bites. Someone gathered some dry grass and built a fire. The Fernuz men placed some *chariks* into the bed of flames and then bit into their heated "food": old sandals. Perhaps this would fill the stomach. The long-coated ones drew out bread and water to consume. Stepan sat there empty-handed, without even a sandal to chew. In normal times, it would have been completely rude for an Armenian not to share. In fact, feeding guests to the point of extreme discomfort was a cultural norm. In this miserable place, all customs were gone.

Around the campfire, the men related stories that would not be forgotten, no matter how much time passed. The new arrivals were from Sivrihisar, a village in central Anatolia. Having similarly fled the caravans, they had been wandering for some time, their journey already barbed with peril. An hour earlier, they explained, an Arab shepherd had chased them. After this revelation, the men decided to rest briefly. A few dozed off. Stepan reclined next to a tall man named Garabed whose face bore a scar. Stepan unfastened the cap of his half-filled water bottle, lifted it up, and drank some.

Just a few sips, he told himself.

At that moment, a loud noise sounded. *Gunfire*. It pierced the calm of the desert and frantically roused the men. A shepherd was standing on a nearby hill, the same one as before. Nearby galloped another Arab on horseback, holding a rifle. Another bullet now burst into the air. Next to Stepan, Garabed screamed as the bullet ripped a hole in his leg. Yet another shot rang out, this time missing its mark. Swiftly, the Armenians jumped to their feet as the mounted man reloaded. Luckily, the gun possessed only one barrel. Fanning out, Stepan's friends advanced on the mounted shooter, stopping and hiding behind stones with each firing. Crouching behind a rock along with the writhing Garabed, Stepan tracked the action of his companions, in awe of their bravery as they steadily gained ground, brazen even though their only arms were arms. Surrounded now, the shepherd began to back away; the triggerman, out of bullets, also hastily retreated.

"Hurry," one of the Armenians said. "Let's get away before the man goes and tells his friend, and they return ready to pursue us."

But Garabed remained splayed out on the ground, bleeding profusely. Feverishly tearing up their rags, Stepan and the others made a tourniquet and placed it around his leg to stanch the flow of blood. Wrapping their arms under his, they pulled him up and then quickly limped forward, a chain of humans, strangers helping one another. For a good while, this worked, but Garabed was growing weaker by the step; he'd lost too much blood.

"Friends." Garabed halted. "I'm no longer able to walk. Leave me. This must have been my destiny. You go ahead. Hurry up and get away. I'm going to sit here."

Listening, the men lost their composure. Stepan didn't want to leave someone in need, but he had no choice. Staying meant they'd all die. Wordlessly, he stooped down and kissed Garabed goodbye, grieving for the man he had just met. Each man took his turn doing the same.

"May God be with you," Garabed cried. "Tell our hellish life to the rest of the world."

Stepan promised him he would, and the band of eleven dwindled to ten. Walking onward in silence, countless more miles passed under their feet.

Two days had passed since Stepan had escaped from his caravan; his bottle felt light. It was time to have some water. Stepan raised the bottle to his cracked and split lips and drank. *Ah.*

Just a couple more sips.

He indulged himself. He would never have enough. By now, the interior of his mouth was unbearably dry, his gums, his throat, the insides of his cheeks. Swallowing was difficult, especially with little saliva. His eyes were steadily sinking into the pools of dark circles he already had. Stepan tilted the bottle completely upside down, but no more water poured out. His two cups were gone. Standing there, with his empty bottle, fear overcame him: *What will I do without water?*

At this juncture, Stepan needed a new plan. He was not yet halfway to his goal. In the middle of this infinite desert, he dug into his pocket and pulled out his crumpled map. Yes, they were all rambling toward the southwest, the same direction Harutiun Atanasian had taken. Already thirsty and fatigued, Stepan knew he couldn't last the six more days needed. There had to be a shorter way. He could live only a few days without water. If he had a steady supply, even without food, he could maybe last a few weeks, wander until he located the Euphrates, provided his feet didn't give out first.

A man his size would have needed around four cups of water a day, twice the amount he was carrying for one week, and that was without extreme exertion, without this scorching heat. Already, his body was trying to conserve fluids. He lost his fluids through perspiration, the way he cooled and regulated his temperature. Stepan needed to drink more fluid than he was losing. Making it to the Euphrates River sooner was their only chance, he decided.

Scrutinizing the paper in his cracked hands, his amateur cartographer's sketch, he remembered his friends from the caravan talking about a different route, straight to the Euphrates River. In all that open space, he had only a vague idea of his whereabouts, the path of the Euphrates marked by a simple squiggle not drawn to scale. He thought about water all the time now. There was another way, he explained to his companions, an alternative passage that would get them to the water in just four more days. At least, that was what he hoped.

Stepan steered the ravaged men into this bleak terrain. They soon encountered five more Armenian drifters. Seamlessly, the way a brook meets a river, they joined together without halting. There was no time. They had reached the desert's deepest depths, where there was no fear of caravan guards. No real landmarks distinguished their trek, and Stepan continued to guess their direction as they traveled by the cool of night, sleeping when the sun was highest in the sky.

Some other escapees in this region told of walking for so long that whole Gardens of Eden appeared before their eyes. The oasis seemed to bloom within reach, a large pool of water framed by trees — or so it seemed.

It was the desert's game, played on all: optical illusions. Occurring at the meeting of refracted light and warm air, it instilled a distrust of one's own eyes. One foreign traveler noted such trickery when passing through this part of Mesopotamia. "Mirage became the ruling factor," he wrote in the *Geographical Journal*. "It reduced the length of [sight] to 2 or 3 miles; any mark at a greater distance was indistinguishable unless considerably raised above the level of the plain."

In the same area as Stepan, another escapee named Dikran Jebejian wrote of finally encountering a well, but the water was too deep for him and his friends to reach. During that weeklong crossing, Dikran also spotted what he thought was a lake in a low field. "Our happiness was indescribable," he wrote in his memoir. "Finishing whatever bit of water that was left, we ran towards the water. What amazement! When we reached that area, the water had dried up, but shined like white water." Five of his friends couldn't bear it anymore, "Why should we take the pain?" they said. "Let us rest here and die looking upon one another."

In the midst of Stepan's water crisis, his bladder became heavy, intense with a familiar urge: he had to urinate.

Oh! he thought excitedly. *I'll fill the bottle.* Steadily, he held the glass and released into it, warming the container to the touch. He set it aside.

By morning, the bottle was chilled, due to the magic of night, his refrigerant. He lifted it up, parted his lips, and drank. Some moisture. This was a fabulous discovery, the wetness the greatest pleasure. It didn't matter the source, or the smell, or the color, which was most likely a rancid amber, not the baby-chick yellow of a hydrated person. As Dikran marched, two of his friends drank from the same fountain. Urine is mostly water, but 5 percent of it is not, and elimination is critical.

With Stepan's thirst hushed, he resumed guiding the men. It was midweek now. Periodically, he would step aside from the others and swallow some more, recycling the harmful substances again. Marching endlessly, he often thought he'd exhausted his strength, but then he would focus on propelling each leg forward, and take another step. Still, the men were demoralized. There was no trace of a river anywhere. A coating of froth now foamed over Stepan's long, thin chapped lips. Born deep inside, it had been growing in breadth, like dough in an oven, until it poured out. So viscous, he could barely breathe.

Briefly, he separated from his friends, carefully rinsed out his mouth, and took two more sips. In this near stupor, the pack occasionally encountered more Armenians; their tattered group had grown to sixty. The crossing seemed impossible. One man fell to the ground, dead. Farther on, they lost another. Every few hours, the motley crew paused to rest. Some men had dates and bread, but a small number were like Stepan — without anything at all.

Five men said they couldn't continue. Stepan watched them not get up again. Five days of walking now, and only eighteen people remained, a worrisome turn, as his "water" dwindled. He had only two fingers' worth left. Too precious to sell to his companion who tried to buy it for two pounds. The man couldn't pee himself, since this, too, had become a privilege. Extreme dehydration steals urine away, along with tears, nature's way of trying to conserve fluids. Still, no crying when it is most warranted; no sweating either, inhibiting the body's ability to cool itself. Blood pressure decreases; the heart rate quickens. Death trails just a few beats behind.

Of this journey, Stepan would remember the particulars of each day, when the sun rose, and when it set. He would remember the faces of those who struggled beside him, the ones who fell and couldn't get up again. The days couldn't be separated according to temperature; the heat was constant, broken only by the precipitous fall of night. A day's rise and fall moving like a chest with breath.

Ahead appeared massive man-made hills. The mounds seemed like something from some bygone civilization, rising ominously from the dusty floor. It was a military structure from a lost age, he thought, from a society that had come and gone. How easily people could disappear. How easily he could disappear. Even in the darkness, they could see that it was a hundred feet wide and towered some thirty feet high. Similar to the others he'd seen, the structures were spaced about two hours' walk apart, resembling tiny earthen hills. Ancient kings must have hidden here, he guessed as the sun ushered in another day, the sixth. Hours later, in the distance, he could hardly believe the sight: Some green hills. Trees too. Were his eyes tricking him? Was this a mirage? No, they were nearing the verdurous banks of the Euphrates River, the great desert behind him. Certain, he drank from his fountain, finishing every last drop.

He climbed to the top of a rise and saw a swish of blue. *Water. Water.* Half an hour more, he assured himself, the proximity exhilarating. Just the anticipation made him feel reborn.

It was midday. He'd made it. Forty-six had not. Only fourteen of them had survived, these cadaverous figures that now closed in on the river. These were the ones, like him, who had had nothing to eat. The ones who ate couldn't pass twenty-four hours without drinking again, their full stomachs haranguing them for more fluids, insatiable in demand. Stepan and the others — the minority — drank fluids only occasionally, yet they could more easily tolerate their thirst. He didn't know the physiological reason for this, how, by fasting, he had urinated less, and his kidneys hadn't had to excrete the food's salts. With fewer bowel movements, too, he would've lost less water. Unknowingly, his misfortune had helped to save him.

Elated by the sliver of water ahead, the withered fourteen quickened the pace, their tired legs drawing them closer. Just a few more moments. *Water.* Then, with a shudder, he noticed some hamlets dotting the river's banks.

"Friends," Stepan said, stopping them, "it wouldn't be smart for us to approach these villages all at once. First, let a few people go, then the others, in small groups. It'll be safer that way."

"Oh ho! You're a coward. We'll all go down together."

"It's up to you," he said.

In that direction, they dragged and stumbled. They were almost at the trees, ten minutes away from the river. Stepan felt delirious with excitement, his throat almost feeling the cool rush of water. *Water. Water.* As Stepan edged closer, his face fell. Several men on horseback were galloping toward them.

The Sandstorm

2007

IT HAPPENED IN A FLASH. After we left al-Busayrah, a whirlwind skated atop the flat dirt and crossed the road in front of us. "Look!" I said. "Yes," Levon explained, "the name is *ajaaj*." As we continued north toward the town of Suwar, another one emerged, spinning like a cartoon Tasmanian Devil. "How cool!" I said, having been familiar only with hot-air blasts from subway grates, not natural weather phenomena. In another second, I realized this "cool" tornado was hurtling toward us. "Oh my God!" I screamed. "Stop the car." But now we were engulfed. Like rain, the ocherous sand pummeled our windows, blinding us. My driver, Antranig, kept his hand steady on the wheel, not slowing. "Stop the car!" I said again. Just as quickly, the *ajaaj* tired of us and moved on. Antranig and Levon glanced at me, then at each other. Both burst out laughing. "It's no problem, Dawn," said Levon, his eyes crinkling.

"You probably see this all the time," I said sheepishly. "*Ayo*," they affirmed in Armenian, still chuckling as we proceeded down the black asphalt road. Every so often, I would remember my new friends in the secret police and peer out from the back seat, checking behind us. Sure enough, there was the station wagon from al-Busayrah. Eventually we crossed a bridge suspended over a tepid Khabur River. It was much smaller than the Euphrates, reminding me of the muddy Sakarya back in Adapazarï. "This is Suwar," Levon said. I glanced at the sandy shores, once a camp, where my grandfather had watched many disappear to the other bank, never to return. Progressing, we drove through its main — and seemingly only — strip and found a small store. It was late afternoon, and I was parched. My bottle of water in the car had become too hot to drink.

The secret police pulled over as well, right behind us. *Why are they called that?* I wondered. There was no effort to hide, only intimidate. New to being stalked, I didn't know how to act, since my mother had somehow skipped that etiquette lesson. *I should be really nice,* I decided. *Maybe then they'll see my intentions are honest.* So I entered the shop and bought us some water and them a few bottles of Coke. "Give this to the men back there," I said to Antranig. The policeman behind the wheel twisted off the soda's cap and smiled. From a distance, I lifted my glass as if in a toast. They raised up their Cokes too, and suddenly I felt like we were in a soda commercial, extolling the virtues of a refreshing drink as sweat dripped off our brows.

Not many other vehicles sped down the parched road; the view was monotonous until a wedding procession of ten cars, festooned in red and white, passed us, honking. I continued snapping photos as we drove my grandfather's last march. Checking on my friends behind us, I noticed them accelerating and veering into the lane of oncoming traffic. Now their car was beside us, so close we could have high-fived. In fact, the driver was waving. "What is happening?" I asked Levon, fearful we were being pulled over.

"They are headed back," he said.

"Oh, they are leaving me now?"

"Yes, I think," he said. I felt so relieved. The sun was beginning to set, the richness of the caramel desert gradually muted. Trucks with Iraqi license plates hustled past us. We kept on, and in the distance I could finally see a huge hill, moonlike, with scalloped edges, the shape of a plateau. The road had been built around it, steering right. The sign said it all: *Mrkada,* or Marqada, as it is alternatively spelled. The river rippled a few hundred yards away, its edges rimmed in green.

We parked in front of the mound. On the slope, a small Armenian church blessed the mass grave. Exiting the car, I could see an old caretaker in traditional garb. Next to him were other men in dark Western clothing. My stomach knotted as I recognized more Mukhabarat. As it turned out, I had crossed from the province of Deir Zor into al-Hasakah, and that apparently warranted a changing of the guard, just as it had for my grandfather. We'd entered someone else's jurisdiction. The caretaker, seemingly unfazed by all the new company, took out his key and opened the door. Inside was a glass casing of bones unearthed from that location. We peered into the aquarium of the dead. A clavicle in one corner, skull fragments, and a femur in another. Small bits of many lives. Were they from Baba's doomed caravan? Was that Yeghisapet Sarian of Izmid? Or the woman Roksanna, who'd given him her bottle of water? My eyes welled up, the tears streaming down my cheeks. For some reason, the bones made the horror more visceral than it had been in al-Busayrah.

Needing some air, I walked out to the front steps and took in the windswept panorama. This place felt eerie, as if the evil had bled into the earth. For as far as I could see, the land was flat. I thought about those scared Armenians, their escapes so easily spotted in the daylight. No wonder Baba chose night as the only exit. This entire day, I had been trying to figure out the location of his escape. He had said it was near a hill. Was this the place? Or farther north in Shedadiye, where we were now going?

The sun was fading. The collage of colors resembled the brushstrokes of an Impressionist painting, only a faint hue separating the sky and the ground. In the distance, oil fields lit the night like stars. The war in Iraq was still raging. By the time we retraced our way down the road, toward Deir Zor, it was pitch-black, the silhouette of Marqada barely visible as we passed. "Pull over, please," I asked Antranig. He complied, and this time there was no one behind us. I exited the car and veered away from the road. I wanted to know what it was like to walk without light, as my grandfather had. The sand blew into my eyes. Tearing up, I made my way forward, toward the Khabur River. Each time I took a step, my foot dropped into a ditch, not visible without a flashlight. Somewhere in this region, my grandfather had fled.

After returning from this trip, I learned the exact location. Following several years of research, I found an article, published in 1940 in Romania. Another survivor of my grandfather's caravan named the spot, this forsaken spot, Marqada. Somehow, I had just known.

PART FOUR

Refuge

Betrayal

1916–1917

THE FIRST THIEVES STRIPPED THE ESCAPEES BARE. The next arrival jabbed a sword into the torso of Stepan's friend. "Stop," Stepan begged as the man probed for swallowed coins. "Stop." Miraculously, the bandit relented, leaving the Armenian with only a bloody gash.

Relieved, the fourteen naked men gathered themselves up and edged toward the Euphrates in search of water, the rushing sound rejuvenating, the scent fresh and crisp. *Water. Water.* More bandits fell upon them. What could they want? There was nothing left; the men were like lemons squeezed down to the rind. Forced onto the dirt, the Armenians sat just several hundred feet away from the Euphrates. Stepan's last sip of water had been four days earlier, and his last food, a handful of raisins, two days before that. The leader now commanded them to open their mouths, and there, tucked amid the molars and the gums of his first victim, he found a coin. Moving down the line, the bandit inspected the next man. Soon, it would be Stepan's turn. With his tongue, he could feel his hidden gold coin that the wealthy Yeghisapet had given him before fleeing the caravan. His sole possession and chance at a future. Ever so surreptitiously, he removed it, and buried it in the sand, beside his foot.

The bandit peered into Stepan's rancid mouth, enduring the blast of ammonia and decay. A tense moment, and then he pulled away; the gang retreated with their meager haul. Stepan retrieved his coin and stumbled the remaining distance to the Euphrates. His feet sank into the marshy embankment; he was really there — the fourteen of them had made it after their weeklong trek. He cupped the water to his mouth, the liquid refreshing as it gushed over his swollen tongue and down his dry throat. A minute later, he drank some more and then had to pause. He couldn't consume it any faster, because his stomach wouldn't allow it. Crouched there, steadily taking in more of this mighty river, he thought about its godlike power. This water, a gravesite to so many Armenians, was now restoring his life.

On its banks, he inhaled deeply, and exhaled, finally able to breathe.

Strangely, the more he drank, the more his appetite washed away. He didn't understand why, he hadn't eaten in ages, yet he felt full. Still, he knew they all needed food — and help. The fourteen men made their way to a nearby settlement that probably housed the thieves who had just robbed them. Walking down the main street without clothes, Stepan wove his hands over his crotch. They came to a square. The wild Armenians sat down and crossed their legs in embarrassment. Men and women wrapped in flowing headscarfs and loose robes converged to witness the spectacle of bare skin and knotted hair, this calligraphy of bones. Still covering their genitals, Stepan and his friends pleaded to the crowd for food. At the hamlet's outskirts more commotion erupted: two mounted gendarmes and a third man galloping toward them. They dismounted, parted the onlookers, and approached the strangers. Stepan's heart almost pounded out of his chest.

"Where are you from?" the third man demanded in Arabic. He was a commissar who had been traveling from town to town alongside the gendarmes, probably to collect taxes for the Ottoman government. "What kind of display is this?"

"We were coming from a nearby village a couple of hours away and we met robbers," said an escapee who spoke Arabic.

No response. Stepan tried his luck with another explanation. "My friend and I were staying with a gardener in Deir Zor and had been on our way to Raqqa," he said as his friend translated. "But as we approached this village, thieves robbed and stripped us."

"Don't feed me lies," the commissar said. "I know where you come from, where you escaped from. Now, what do you want me to do with you?"

"It's up to you, Bey Efendi. Do what you please," Stepan said. Perhaps by cooperating he would ingratiate himself with the man — and gain some time to think.

"Where would you like to go?" he asked.

"To Raqqa," he said, the safe haven.

The official now switched to Ottoman Turkish so the onlookers would not understand. "These people are savages. If I set you free, they'll kill you. We're visiting villages on our way to Raqqa. We'll take you with us; that way you'll be safe."

"May God grant you a long life, Bey Efendi," the poor Armenians chorused.

The commissar instructed his colleagues to distribute the Armenians among the Arab locals for the night. "Tomorrow we'll take them with us."

At first light, Stepan and his friends happily joined their escort westward. In mid-October 1916, the nights shivered frigid, and the days were steamy with heat. Stepan searched the road for fabric scraps to protect himself against the elements and hide his shame. Scavenging strips of straw and rags, Stepan and the men tied them together around their waists like an absurd troupe of hula dancers.

Down the Euphrates's left bank, the group walked westward, stopping at small settlements where locals offered shelter and food. After four days, they finally neared Raqqa, halfway between Aleppo and Deir Zor. Since they were already on the same side of the Euphrates, they didn't have to cross the river. The last time Stepan had been there, some five months earlier, he'd seen Khoren, written to his family, and felt all was possible. At last, he had that same feeling of freedom again.

Then one of the escorts barked new orders, "Form a column in twos." He led them to the promised land — then straight into jail. In the cell, there was no roof overhead, only the open sky. Lying on the stone floor, the grown men wailed as the new guards kept a close eye. All that hope, all those plans for reuniting with their families and friends, was gone. Hours passed, and the door swung open to reveal a sergeant.

"Yes, of course, everything has its turn! Who told you to join with the Russians against us?" he interrogated them. "This is what happens!"

Fear seized Stepan with one hard slap. There it was again, the propaganda: that the Armenians had sided with the Russians and rebelled against the Ottomans. Stepan had heard this before—all the Armenians and Turks had. Whatever Stepan and his friends said to the contrary—that they were nowhere near the Caucasus front—the sergeant wouldn't have it. The next morning, two sentries converged, shouting, "*Kalkin bakalim!*" Get up!

"What do we have to do?" Stepan asked.

"*Deir Zor'a gidejeksiniz.*" His response nearly sent Stepan to his knees. They were going back to Deir Zor.

More upsetting news followed: Zeki Bey, the governor, had personally ordered their transfer. That meant they'd be dead in a week. The guards escorted the fugitives across the Euphrates onto the right bank, and handed them over to a new company on horses. Stepan knew they didn't have much time. "Our end is death," he whispered. "Come on, let's jump the guards; there's fourteen of us and only two of them. We'll take their guns and flee.

"Even if four of us are killed, ten will survive. I'll be one of the first to attack and probably the first to be killed, but I'd be happy if I could liberate my friends."

Five or six agreed. The rest, however, did not. "The outcome will be worse," one insisted, "they'll take it out on our people." "We're not strong enough," protested another. Without everyone's cooperation, Stepan knew they couldn't escape. Demoralized, he walked eastward for the next hour and a half in a near stupor until another settlement appeared, dimpled with tents. Under guard, Stepan and the others camped on the outskirts and then set out again at daybreak along the river's bluffs, not halting until Sabkha. There, sixty-five miles before

Deir Zor, the guards threw them into a decrepit cell. This time there was a roof, but no windows. Only a dark prism of walls that revealed nothing of the shifting war beyond them.

A world away, the U.S. Congress declared October 21 and 22, 1916, "Armenian and Syrian Relief Days." It was an effort to raise money for the Armenians "in view of the misery, wretchedness and hardships, which these people are suffering."

While some provisions were making it to the starving Armenians, distribution was still being thwarted. For many, however, it was too late. Masses were already dead, wrote Secretary of State Robert Lansing, their deaths the result of a "studied intention" by the Ottoman government. "The true facts if publicly known would shock the whole civilized world," he wrote to the American embassy in Berlin.

The German government was also aware of the unfolding carnage. New intelligence informed the Germans about a wave of killings along the Khabur River, where Stepan had just escaped. "As far as we know, the annihilation campaign was implemented by the Mutasarrif, Zeki Bey," wrote the German consul of Aleppo, Walter Rössler, to Germany's chancellor. "There is no doubt now that hundreds of thousands were sent to the area around the Euphrates River." Some survivors had begun to relate their harrowing stories of what had transpired in Maraat and Suwar, and Rössler enclosed their accounts. In a caravan of seventeen hundred, thirty-one survived by hiding under a mound of corpses. Another survivor described two hundred gendarmes surrounding their caravan and then opening fire. Yet another testified to a scene hauntingly similar to what Stepan endured, with the massacres unfolding near a hill.

Over a two-month period in that region, tens of thousands were murdered. As a whole, the Armenian deaths might be as high as one and a half million, according to a letter from the new German ambassador in Constantinople. Yet the Germans were hesitant to provide aid. Certainly, the Ottomans, their ally in the Central Powers, would take it as "interference in the inner relations of the country," explained the German foreign minister Arthur Zimmermann. The Turks had pressured

Germany to recall its ambassador to Constantinople, Paul Wolff Metternich, for this very intrusion.

In the tomblike cell, Stepan was starving. Several days had passed without food, and the smell of the guards' meals, held on the door's other side, taunted Stepan and his friends. The men had tried to quell their hunger by sucking on the cowhide strewn on the floor, the taste akin to leather shoes. Through the cracks in the entry, the prisoners stared at the sentries gnawing on lamb chops and other meats and begged them to throw them the bones. On rare occasions, they received the scraps and their watermelon rinds, and they ate every last morsel as if it were a feast.

In the coming days, the sentries deposited ten more Armenians into their cell. Stepan was allowed just one visit to the toilet every twenty-four hours, and his bladder felt as though it were ready to explode. Having used his one gold coin as a bribe for bread, he couldn't buy his way out of the misery, out of this slow death. Unable to stand this anymore, he addressed the two dozen men around him and proposed writing a letter to the *müdür* of Sabkha requesting bread. "If not, kill us." The prisoners agreed. One of the newest detainees, a man from Marash, spoke eloquent Ottoman Turkish, so he penned the request. Fifteen minutes after summoning the *müdür*, the door opened. The director stood there studying the caged men, chiseled down to the bone with starvation. Stepan stretched out his arm and gave him the letter; the sharp outlines of his wrist bones were clearly visible. As the official read, Stepan tracked his eyes.

"Well, my boys, I'll bring it for you," he said now. The *müdür* turned away, wiping his eyes, trying to hide his emotions. *Lavash* arrived not long after, one apiece; the *müdür* had kept his word. The men pounced on the flatbread. The next week, however, the food did not come, nor the following week. Only on the morning of the final transfer did the *müdür* appear with half a piece of bread per person.

"Farewell," he said. "My boys, do not blame me. It's not in my hands; I must obey the orders I receive. They're demanding your return to Deir Zor and you have to go. May God's mercy be on you. Go in peace."

Once more, the men shadowed the Euphrates until their feet gave out. In tiny Ma'din town, they were jailed again so they could rest before completing the march to Deir Zor. Stepan glanced around his new cell, bright from sunlight streaming through three windows. He noticed something unusual — the door was guarded but the windows were not. A way out, he realized. After explaining this to the others, he volunteered to bear the most risk and go first, given he was the youngest and most agile.

That night, he hoisted himself onto the windowsill and carefully started to lower himself down the wall's other side. He reached his leg down, then recoiled suddenly, detecting something soft underfoot: a watchman, fast asleep. Alarmed, he examined the next window and the one beside that. All were guarded, unlike the previous evening. Stepan slid back inside the cell and slumped against the wall next to the others, defeated, and eventually drifted to sleep.

The next morning, the guards smirked. "You were going to escape. Why didn't you?"

"No," Stepan lied. "We had no intention of escaping."

Stepan soon heard that two of his cellmates had betrayed him during their bathroom break. Too sick to escape, the men didn't want to be left behind. Now they were rewarded for their loyalty to the guards: they were spared from the transfer. Stepan didn't know whom to trust anymore and resolved to escape — on his own.

The detainees struggled on the road, under careful watch. One gendarme was stationed at the front, another in the middle, the last at the rear. All were riding horses and outfitted with weapons. Stepan searched for an escape route as the path weaved between the river and the bluffs. Before long, they reached the next village, Tibni, only twenty-eight miles from Deir Zor. The distance was contracting like a noose around his neck. He could almost feel it. He was panicking. An open field stretched as far as he could see. He thought of animals, how they bucked and bit in order to avoid death, how they put up a fight. But his faith in his own luck was gone: *Let them shoot me. One bullet is enough to put an end to this life. I can't take it anymore.*

The terror was choking him. He couldn't return to that cursed hill where his convoy had been massacred. He, who had endured everything, had reached his limit. He pictured his family: his siblings, Zaruhi, Armenag, Mari, Arshaluys, and Aghavni, and, of course, his *mayr*. He could see them so clearly, those long noses and hooded eyes. "Goodbye." He pressed on, sobbing, his feet numb, his mind and eyes oblivious to the time and place. "I shall die either in Deir Zor or here," he told himself. *God, God, forgive me!*

Just then, he heard a voice, its origin unknown. Was it his family? His conscience? Or God? *Don't sleep. Keep your eyes open. Look around.* It was as if someone had shaken him awake. The convoy had veered off the main road and onto a shortcut, a footpath that sliced through brush. It was less than an hour to Deir Zor. Stepan studied the guards. The one at the front gazed ahead. He whipped his head around and scouted the side and the rear. Some luck; the other two had dismounted and were squatting down, defecating.

Now's my chance. Be brave. Summoning all his courage, he stepped off the route. Standing twenty feet away, he placed his hands in front of his crotch, pretended to relieve himself. Surprisingly, the other prisoners didn't notice. Their heads only pointed forward, their bodies buckled and weary. At this point, the group had advanced two hundred more feet. Stepan dropped to the ground, his chest and belly pressed down like a snake. He froze in the low bushes, wild all around. He knew that the indisposed guards were still busy with their bowels, and had not yet ridden through that area. From horseback, he would be easy to spot. He listened for the beat of hooves. *Clip-clop. Clip. Clop.* They were drawing closer. Now they were there. *Clip. Clop. Clip-clop.* Wait — they were passing him. *God's hand is in this.* Still, he was in danger. Once they caught up to the group, the gendarmes would notice him absent from their two dozen captives. He crawled and scooted until he came to a hill. Hallelujah; the sharp rock bore a slender opening. He rose to his feet, angled himself, and tried to squeeze in, but the aperture was too narrow. He pushed some more. Success! He was now hidden. He touched his hand, felt a sticky substance. Blood, he realized. How had that happened? He hadn't felt any pain.

He stood still, his ears awake to all sounds, especially the slap of footsteps. But there were none, not now, not later. He must be alone. Only the bone-chilling desert for company. Upright in that awkward position, he waited for night to fall.

The day was waning; hours had passed since his escape. He could scarcely move his body in this crevice. Why had he been able to slip away, the others not? A divine miracle had taken place, he believed. But now what? He began to pray.

O Lord, You saved me again. You've done so much for me. Now, please send Your guardian angel to guide me, to show me my way.

The filtered light ebbed until there was none. He wedged himself out of the cleft and emerged to darkness, but his way was lit by brilliant stars. The barren land felt muffled in silence. To orient himself, he located the Euphrates and stepped ghostlike onto the path toward Raqqa and away from Deir Zor. Despite his recent incarceration in that town, it was the only place that offered any type of refuge. Every so often, he would dive down flat, press his ear to the ground, and listen for the slightest rustle or vibration of horses' hooves. If he heard people coming, he would immediately slide into the brush and remain there until they passed, as still as a reptile.

Sometimes the vibration was caused by gendarmes, other times by armed Arabs on horseback. It was about the middle of November 1916, two years into the war. Stepan watched the traffic surge past as he had during all those months in the roadside encampments, not always knowing its relation to the greater conflict. At the moment, the war was moving to the Sinai region, where the British had amassed a large number of troops. Also, the Arab revolt against the Ottoman government hadn't turned out to be as big as expected, despite the capture of some key towns.

But that would change with the arrival of a young British officer who would ally with Faisal ibn al-Husayn, the Arabian helping to lead the uprising. His name was T. E. Lawrence, soon to be known simply as "Lawrence of Arabia."

In the hours after Stepan fled the guards, he passed Tibni and the ruins of Zelebiye and Halabiye, ancient settlements. Just before dawn, he reached Ma'din, where he'd attempted to climb out the prison window.

The cold air needled his skin. He yearned for shelter. He veered toward the riverbank and walked for another hour as the daylight turned the silver waters to blue. He stooped to drink and noticed the caked blood on his body, cuts and scrapes from the crevice. He washed himself and rested. For fifteen hours, he had been on foot.

Where should I go? Which way? Raqqa was the natural answer, but it lay on the Euphrates's other bank, more west. He had the skill to swim across, but not the stamina. There, on the shore, he noticed some tents, and he moved closer to see an Arab couple huddled at a campfire. They spotted him. Stepan knew he looked strange. His body was uncovered save for his rag skirt, his hair that of a beast. *"Gel, gel,"* the Arab man said. Come, come. Now he said it again: *"Gel, gel."* Tentatively, Stepan edged toward them, near enough to feel the heat of the flames. "Add more wood to the fire," the Arab man instructed his wife. Stepan began to gesture that he was hungry. The man uttered something indecipherable to his wife, and from her tent, the matron retrieved a big pot of *khavourma*, a dish made from chunks of lamb that Stepan used to enjoy with his family. With two hands, the woman presented it to Stepan, signaling him to eat. At her urging, he bit down on the meat, the animal flesh, and then moved on to some bread, which they warmed just for him. He gorged until he couldn't take another bite. Feeling less lightheaded, he approached the nearby cluster of pitched tents, mixing pantomime with Arabic words to convey his desire to work.

At last a man answered him: "Can you work as a shepherd?"

"In my country, I was a shepherd," he lied.

"Sit down." The man studied him before speaking again. "What is your name?"

"Misak," he replied.

"From now on, your name won't be Misak," the stranger said in Arabic.

"Then what?" Stepan asked.

"Mustafa."

From his tent, the man brought out a *mahrama* and a flowing *zboun*, a long-sleeved shirt that extended past the knees. "Throw away the rags you're wearing around your waist," the stranger said. Stepan inserted his head through the shirt's opening. "Come, let me shave you like an Arab so nobody can recognize you." The *emanetji's* beard was forested with thick tendrils that curled down his neck; only the top of his cheeks showed. The man took a sharp blade and slid it along Stepan's neck, around his Adam's apple, and across his cheeks, hirsute with growth. His hair fell away in clumps. Next, the Arab took a cord and looped it around Stepan's waist, so slender now. Then, taking a *mahrama,* a piece of cloth, he adorned Stepan's head, the white fabric draping past his hair and ears and shoulders. On top, he crowned it with an *agaal,* a black ring. He handed Stepan a mirror. The *emanetji* hardly recognized himself. *I look exactly like an Arab,* he thought, and he thanked God for the re-incarnation.

This respite was short-lived. The man's wife ran up to Stepan. "Follow me," she implored, and she led him to a tall pile of straw. "The gendarmes are coming. Get under the straw. They must not see you." Into the soft stack, Stepan dove, and the woman hastily threw fistfuls of scratchy stalks over him. Only his eyes remained visible. He tried to quiet his breath and any other sound. From between the crosshatchings, he watched the drama unfold: two gendarmes had ridden in on horses with a passel of twenty Armenians.

"Are there any Armenians?" they asked.

"No," the couple replied.

Next, the gendarmes approached the *muhtar,* alderman, and he had the same response; he even swore by his beard. Convinced, the gendarmes left.

"Come out," the Arab's wife called him as he trembled.

That night, for the first time in two years, Stepan fell asleep feeling both full and safe. The next day, when he tried to find work, the Arab man said, "You rest one more day." Stepan couldn't argue, given his feeble state, but he was also relieved because he'd lied about his experience.

What am I supposed to do when they trust me with a few sheep or cows? Still, he had bigger worries. The gendarmes were scouring these roadside villages for his kind, corralling and driving them toward Deir Zor. The generous couple always warned him. They knew an hour in advance, the flat ground perfect for spotting oncoming traffic. Even though Stepan had a disguise, he was living in fear, given the hamlet's proximity to the road. He couldn't stay.

At midnight, Stepan removed his *zboun* and *mahrama,* not wanting the kind Arabs to mistake him for a thief. Back in his rags again, he slipped off unnoticed, eventually encountering another settlement away from the main road. On the outskirts, he fell into a pile of wheat chaff. The next morning, horses neighed him awake. During his search for work, he encountered more gendarmes and convoys. Always, he changed direction. His hair and shave appeared native to this desert, but he couldn't speak the local language. In four days, he had some luck. After Stepan haltingly introduced himself as Mustafa, a man and his family hired him as a shepherd.

"Very good," the man said. "I need someone like you. Come sit."

First they ate. Someone brought a basin of water, and, one at a time, they all rinsed their hands. Then Stepan joined them on the ground, legs crossed, the Arab way. A large pot was being filled with *tarhana,* a soup made of dried bean curd. He watched the people deftly cup their hands to eat. Each time Stepan tried to mimic the gesture, the watery dish leaked from his fingers, and he barely managed to swallow a mouthful. Sensing his difficulty, someone retrieved a porcelain cup. Stepan submerged it so he could finally consume his share, one cupful, sipped like tea. It wasn't nearly enough to tame his hunger, but dinner had ended, and the family was gathered around a fire, warming themselves as the night grew late.

"Come, I'll show you where you'll sleep," the man's wife said.

Stepan trailed her to the tent's other flap, where the animals lived. She handed him a dirty quilt and pointed to the ground. His bed. He folded the quilt in half and slipped between the two layers, exhausted, then rolled up a donkey's blanket for a pillow. He was safe. Immediately, his body began to prickle. Fleas jumped around everywhere — on his blankets, legs, and arms. He swatted and chased their jumps with his fingers. Under siege from these wingless creatures, he couldn't sleep the entire night, itching and twitching.

The next morning, he washed his face, dried it with the end of his shirt, and joined the others around the fire. He waited for breakfast, but there was none. After Stepan rested for several days, the woman appeared, leading half a dozen cows, including three calves. She gave him a mangy *mshllah*, an Arab coat, and asked him to follow her.

"The animals know their way to the Euphrates," she said as they walked. "At noon they go to drink water and then graze along the river until evening. They return home late. It's your job to keep an eye on them. At noon, when they go to the river to drink, they pass by in front of our tent. You then drop in to get your bread."

Fifteen minutes later, the path opened to a pasture. She handed him a staff. Stepan didn't know how to shepherd animals, but he learned quickly by observing the other herdsmen, who soon became his friends. Each day, he'd pick up his bread at noon then trail his herd to the riverbank, where he laid down. The job required his constant attention, a welcome distraction from the agony in his heart. He was relatively safe, he knew, and only hungry, not starving. But still, he felt as empty as before. All he wanted was to see his family again.

With the passing weeks, Stepan's portion of *tarhana* soup steadily decreased as the woman served him less. Soon the contents of his cup resembled a summertime riverbed — almost all dried up. At night in the stable, he listened to his stomach grumble as the family conversed deep into the hours. He could hear the murmurs of the parents as well as their three children. After sunset, one of their daughters, the wedded Fatma, would make love with a man who was not her husband, the affair visible to Stepan from his flea-ridden bed. Normally the scene would titillate, but he felt worlds away from carnal pursuits. Fatma's single sister, Aliye, also yearned for romance and regularly flirted with Stepan, trying to seduce him.

Confused by his rebuffs, she asked, "What are you? Aren't you young?"

"I am, but I'm very weak," he said. If the war and relocation had not overturned his life, he would probably be married by this point, with a child. Now he didn't know if he'd ever have a family of his own, ever enjoy the touch of a woman again. He was still alive, yes, but incapable of enjoying pleasure. Instead, he focused on shepherding and contemplating the changing tides of the river, and the war. In his dreams, in the distance, he saw his home — or so he hoped.

It was nearing the end of 1916, and Arab tribesmen were continuing to resist the Turkish troops. More news of Armenian massacres appeared in the press; Americans reelected President Woodrow Wilson on his platform of staying out of the world war. In December of 1916, the Battle of Verdun, on the western front between German and French armies, finally came to a close. No one had won, all had lost, with seven hundred thousand casualties. The Germans defeated the Allied Powers in the Battle of the Somme, and it came at an even higher cost for both sides, with the dead or injured exceeding one million. And still the war droned on into 1917.

That winter with his sheep, the hours passed ever so slowly. Beside the Euphrates, Stepan would remove his one ratty shirt, which was crawling with fleas, and lie on the ground naked. He'd pluck up a black insect, crush it between his fingers, then snap it in half. It became like a sport, as he began to count, his record for one day's kills six hundred. To while away the time, he'd also play games with the shepherd boys, as he was doing this particular morning when two gendarmes passed on horseback in the distance. Noticing them, the boys signaled and yelled, "There's an Armenian here. There's an Armenian."

"Shut up, boys, shut up!" Stepan lashed in broken Arabic, their betrayal catching him by surprise.

"There's an Armenian here, there's an Armenian!" the boys screamed louder, almost reveling in their power.

The gendarmes turned in their direction. Stepan knew he was finished. He had no chance of escape. The turn from safety to danger was dizzying.

The gendarmes rode toward him. Not again. The gendarmes addressed him, and his stuttered reply, in broken Arabic, unraveled his disguise.

"You look like a real Arab," a gendarme said with disgust. "Who knows how many gendarmes you've deceived. Now you're trapped. Get in front of us, *giaour*, and walk!"

He jettisoned his staff, and a stab of pain pierced his back. From atop their horses, the soldiers prodded him toward the main thorough-fare with the butts of their rifles. He knew the word the gendarmes called him, *giaour*. This was what the Muslim Turks called Christian Armenians like him. The horses began to trot quickly and Stepan couldn't keep up. Weak and hungry, he labored forward anyway, knowing how this would end. The road to this Mesopotamian desert was long. Already, Stepan had lived more days than he'd thought possible.

"At the pace he's going, we'll be late," the guard declared. "Let's kill him and throw him into the river. He'll be rid of us and we'll be rid of him."

It was all happening so quickly, Stepan didn't have time to think. They walked off the path back toward the Euphrates River. Desert stretched as far as he could see. There would be no eyewitnesses. After all this time, his luck had finally run out. They commanded him to the water's edge. After they killed him, his heavy frame would drop into the current. Stepan had seen it before, the bodies drifting away.

"I can't be troubled with you all the way to Deir Zor," the policeman said. "In fact, if you go there, you'll also die." He loaded his rifle.

Stepan's face was smeared with tears, and his lips parted almost on their own. "For the love of God, forgive me," he said in Ottoman Turkish, loud enough for them to hear, and stepped toward them a few inches. "Please give me five minutes to say my prayers, then shoot me, so the Lord condemns neither me nor you."

They agreed. Raising his long arms above his head as if to summon a higher power, Stepan wailed. Between sobs, he began his lament, which sprung from his mouth without plan or thought, "Dear God, my Father, You know that I have four children, the oldest is six. We were not rich; my wife was always sick and could not work. What will happen to my children? They'll cry every day for a piece of bread. I leave it up to You."

The chilly river splashed around him as he paced back and forth. "If You want me to die — if it's my destiny — I'm ready to die. I have no time to repent my sins: forgive my sins. Forgive also these people; they are kindly, good people like me who have children of their own, but they are under orders. They are obliged to carry out what they have been ordered to do."

Now he turned to the gendarmes, and more words suddenly came to him. "I'll call out to God three times. The third time I pronounce the name of God, shoot me.

"God . . ." he said, and he dropped to the ground, the whirl of the rapids and his breath the only sounds.

"*God . . .*

"*God,*" he said one final time, and he braced himself for the firing squad.

Three of Stepan's sisters and nieces. Top row: Araxi, Mari, and Louise. Lower row: Arshaluys, Madeline, and Aghavni. This was taken around 1921, the year they left Constantinople and immigrated to the United States, laying the groundwork for Stepan and his mother to follow.

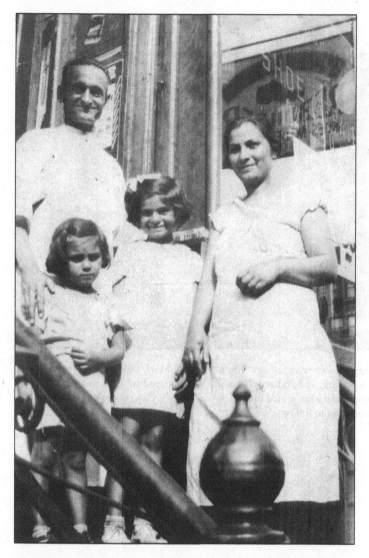

"God Bless America": Stepan and his family finally arrived in the United States; here, they stand near their home in Spanish Harlem in 1931. From left: Stepan, Anahid, Alice, and wife, Arshaluys.

Newly widowed, the eighty-five-year-old Stepan stands beside his daughter, Anahid, and granddaughter, Dawn, in 1971.

A typical street in Chai, Turkey. Not far away from here — at least in terms of mileage — sprawled the internment camp the Miskjian family was deported to in 1915. Alongside thousands of others, they lived in makeshift tents beside the train station.

Down the street from the store in Adana, Turkey, where Dawn had her headscarf makeover.

The caretaker of the genocide memorial in Marqada, Syria, unlocks the door. The chapel was built beside the hill where Stepan's caravan was massacred in 1916. He escaped by crawling on his hands and knees and traveling across the desert, with almost no water, for the next six days.

The village in Syria that is home to the descendants of Sheikh Hammud al-Aekleh, who saved Stepan after the massacres and became like a father to him. Finding the family in 2007, Dawn had the opportunity to thank them for saving her grandfather's life.

Dawn and descendants of Sheikh Hammud al-Aekleh enter the cemetery in Syria, where the revered chieftain is buried. Also pictured: the Bedouin envoy from Raqqa who helped to locate al-Aekleh's family.

At the cemetery in Syria, Dawn pays her respects to a relative of Sheikh Hammud al-Aekleh in 2007.

On April 24, 2015, the hundredth anniversary of the Armenian genocide, Dawn and her mother, Anahid (pictured), marched alongside an estimated 160,000 others to raise awareness of the killings. It was one of the largest demonstrations in Los Angeles history.

The Church

2007

THE ELEVATOR DOORS OPENED to a decadent hotel lobby in Deir Zor, graced with golden columns. As I rushed out onto the freshly polished floors, I could see Levon and Antranig sitting patiently, even though I was late. "I'm so sorry," I exclaimed. "So sorry!"

Turning past some indoor plants, I realized my translator and driver weren't the only ones waiting for me. In another chair sat a policeman from Suwar, one of the men I'd treated to a Coke. He and another officer were glaring at me. Clearly, my soda-pop diplomacy had failed. After leaving Marqada and returning to the Deir Zor hotel I was using as a base, I must have re-entered their territory. This time, I politely waved but didn't smile, feeling exasperated by the continued surveillance.

With the policemen following us out, I glanced at Levon and Antranig, but they didn't comment. As usual, they seemed used to such occurrences, and I had no idea whether the additional scrutiny was worrisome or routine. Pushing it out of my mind, I focused on my destination: the Holy Martyrs Armenian Church in the city's center. Through the windows, I gazed out at the magnificent Euphrates River, its banks thicketed with a jungle that momentarily pushed back the cracked earth. Across the way, plain cement buildings clustered around trafficked streets with long lines of yellow taxis slogging forward, men in robes streaming into shops, mothers holding children's hands. Somehow, when I'd arrived several days earlier, I hadn't expected a bustling metropolis with its own airport. I'd imagined a time capsule, like the dusty outpost my grandfather had seen, a freeze-frame of his caravan's last stop.

We swerved around the traffic circle, where two giant billboards of President Bashar al-Assad stared out. I didn't require any more reminders of the ruling regime — the trailing station wagon was enough. Antranig pulled over near a grand limestone structure, with an arched doorway. Immediately, I recognized it as the Holy Martyrs Armenian Church, a memorial to the Armenian genocide that was built in 1991.

A handful of steps led into the courtyard, built around a pointed chapel that held a museum and the remains of the dead. As I stepped across to it, I could hear my footfalls on the ground; the place was nearly deserted. Much different than every April 24, the genocide's remembrance day, when, I was told, this space filled with busloads of Armenians from around the globe making the same pilgrimage as me: coming to remember, to never forget. The surrounding three walls each had a special meaning, I'd learned. The one straight ahead, with the eagles, was dedicated to the Armenians who had resisted the Turks; the one behind me, with the cross and two doves, was devoted to peace. My eyes were drawn, however, to the friendship wall to my right, which stretched the length of the place and had two fountains in front. It was for the Arabs who had saved so many Armenian lives during the massacres. In Syria, where the small Armenian community had fairly good relations with the government, I felt free to talk about my ethnicity and the genocide. In fact, because of the country's strained relationship with the United States, I had been emphasizing my Armenian heritage instead of my American one — the opposite of my strategy in Turkey.

Descending the chapel's narrow stairs to the remains, I remembered what the building's architect once told me about the sensation he hoped to evoke. "It's like going to be killed," said Sarkis Balmanoukian, a Syrian-Armenian now living in California. "It symbolizes the caves where Armenians were filled and burned."

The balmy air in the courtyard followed me down to the lower level, which flickered with candles casting a warm yellow light. I lit one in the name of my grandfather and all the others who had suffered, and then I turned to face the remains that were positioned around an obelisk topped with a cross. Under glass, amid all the fragments, rested a human skull, nearly perfectly preserved except for a break at the eye socket.

"I didn't want these bones, but the people wanted them," Sarkis had said. "I didn't want the feeling that they are victims only. Usually in memorials, they don't put skeletons or bones, but in the Middle East, they do."

Since entering the church, I'd grown increasingly solemn. The honking of cars from outside fell away, silenced by more of our footsteps. The next room featured photos of starving women and children, a museumlike exhibit of the dead.

I walked up the stairs and out to the middle of the courtyard, where the sun warmed my skin. I stared off, remembering that Sarkis had wanted the feeling of resurrection, of rebirth, after the visitor ascended from the bottom floor. My skin was growing hotter, in that moment before a burn, in the moment that seems to jolt one into feeling alive. I felt truly serene, almost still in mind and body. I'd never been really moved at memorials, but this place seemed different. There was indeed a peacefulness within these walls. I thought about those who were lost, whose bones were here, and I thought about the incredible strength of those who'd survived, who'd stood up, taken another step, and lived to rebuild their community. Sarkis was right. They weren't victims. My grandfather wasn't one either.

Standing there, I heard the rising volume of Armenian voices. More visitors had entered the church. I exited onto the street, and the cacophony of real life assaulted me again, the scream of car brakes. The tranquility was gone, and across the road, the station wagon still waited for me.

Undeterred, we pressed on to the town's old bazaar, where my hungry grandfather had once come to stare at the food. Walking down the teardrop-shaped corridor, I passed stalls of fuzzy sheepskins, rolls of coiled fabric, mint-green and coral dresses, almost anything one could need. The marketplace spilled out onto the neighboring streets, where chain-smoking men sold cucumbers, tomatoes, and scoopfuls of green olives to women toting plastic bags. Soon Levon pointed out an old sand-colored building with green trim around the entryway. It was the old Ottoman office of the gendarmes, he explained. I thought about all the orders that must have been sent through that door, the ones condemning so many lives. We kept on and walked to a quaint riverside café. On the outdoor patio, I gazed at the placid waterway, where people strolled across a pedestrian bridge.

While we were eating our *meze*, I noticed a tiny brown monkey scaling the lines high above, his long limbs moving against a powder-blue sky, his white back turned to us. I quickly finished my bite and hurried to the bridge's entry, captivated by the animal's motion. He was so nimble, neither aware of his growing audience below, nor fearful of a slip. I wanted to get closer, and I strolled across the bridge. Stopping in the middle, I gazed across the river and down the other bank, where my grandfather's camp had once sprawled. Minarets peeked out between sporadic houses and businesses and a tract of empty space. With the monkey at the top now, still undaunted, I walked to the bridge's end and planted my feet, relieved to be on solid ground again.

The Sheikh

1917

STEPAN LAY ON THE GROUND, his breath ragged. Several moments had elapsed since he had pleaded for mercy; the churn of the Euphrates was loud, as if ready to whisk him away. He swung his head toward the two gendarmes. Unbelievably, they were just standing on the riverbank, frozen, not even shuffling their feet.

"Shoot me!" he wailed. "Shoot me!" The river swirled and splashed in its race downstream. More seconds passed. The gendarmes did not stir. Their rifles remained suspended in the air, aimed at him.

One lowered his weapon and reached for the other gendarme's hand. "No, we won't shoot him."

"Yes, we will," the other man countered and continued leveling his rifle at Stepan.

"Let him go for my sake," pleaded the first.

"What's happened to you?"

"This time, my heart wants me to save."

The argument transpired so rapidly that Stepan could scarcely follow the words as the sympathetic gendarme stepped forward and blocked the barrel of the gun with his own chest. The other gendarme pushed, and he pushed back, and the two began to struggle for control of the weapon.

"Run! And pray for me," the merciful soldier shouted to Stepan. "It's for you that I am fighting!"

Stepan leaped to his feet and took off. "May the Lord give you long life," he screamed. His *dua*, his prayer. Even as he fled, he was astounded at the turn of events. Spinning a yarn out of nothing had always been his talent, and persuading others of his sincerity had helped him survive. But this — where had that tall tale come from? How had he spontaneously invented a family drama that compelled a Turkish gendarme to spare his life? And what about that kind Turkish gendarme? Stepan hadn't considered the ramifications of his own act of self-preservation. *He might be dead by now*, Stepan thought as he sprinted back toward the village. That man had been willing to sacrifice his life for his, for an Armenian's.

That night as the rest of the village slept, unaware of Stepan's narrow escape, he went back to the Euphrates River, his mind racing, his stomach aching with hunger. *God gave me my life in the place of bread.* Another miracle had just occurred. Sometimes, God changed the circumstances without explanation. No other reason made sense. Why else had Stepan evaded imminent death so many times?

He stared at the river's inky surface. The waves were crashing against the bank with force, spraying everything with mist, before churning onward. He narrowed his eyes, certain he understood the message in its depths. He, too, would keep moving, like this great water source. He'd never have to see those shepherds who had betrayed him again. Returning to his tent, he started to plot. The next day, he would cross the Euphrates and reach the left bank — the side Raqqa was on, a place free from massacres. Though the waterline was usually low at this time of year, there was no bridge across, and the winter temperatures chilled his skin. He wouldn't risk drowning by swimming. Instead, he decided to construct a boat. Traditionally, locals built a *kelek* from the skins of animals, sometimes using up to eight hundred skins for each. When inflated with air, they transformed into a balloon raft. Seemingly simple, except that Stepan did not have any such materials. No goatskins, no wooden poles to fasten atop, no oars to navigate the current.

After waking up early the following morning, he led his cows to pasture and then walked to the corn stalks growing along the banks. With the ears already harvested and the scarecrow-like leaves eaten by the cows, this was all that was left of the season. He tugged at the thick stalks, ripped up the roots, and piled the materials in a ditch, the base of his clandestine boat-building operation. Over the next few days, he did this until he had gathered enough for his construction. Always he returned to his employers in time for diluted soup.

One morning, Stepan hid a rope under his baggy *zboun* and guided the cows to the river. He waited until evening for the foot traffic to pass, then laid out the stalks, side by side, and tied them with rope. Holding the contraption, he trudged fifteen minutes up the river to a secluded spot, lowered his homemade raft into the water, and climbed onboard. Sinking a little, the boat rocked under his weight. But he was floating. Stepan pulled the raft back and undressed, removing his *zboun* and his *mahrama* and knotting them around his neck to keep them dry. This was it, his only chance of escape. He launched the raft again. He waded deeper, and the river rose to his bare torso, then his shoulders. A frigid wind was blowing, the water like an ice bath. When he reached the middle of the river, his feet could no longer touch the bottom. Stepan hoisted himself onto the stalks, lay down, and paddled with his hands. Waves came crashing over him; the river flowed fiercely, forcing him downstream. He was three-quarters of the way to the opposite bank. He kept paddling, but the current had more might. Exhausted and freezing, he lifted his arms, and his craft zipped along with the drift. For what felt like an hour, he coasted on his jury-rigged raft, speeding past the cornfields. He was wet, trembling. The float bucked and tossed, nearly throwing him off. He gripped it tighter. Finally, near the bank, he let go of the raft and dove in. The river claimed the boat, and he swam to the shore — the shore he had jumped from. Not only was he on the same side, but he was still close to the village he'd tried to escape. He felt cursed. In the distance, the cows were shuffling their hooves toward home. Drenched and shivering, he dashed to catch up and chaperoned them the rest of the way.

"What happened to you?" the peasant family asked.

"I went into the river to wash myself."

"Are you crazy?" one asked. "To do such a thing at this time of year?"

He wasn't sorry. He was determined to leave, to find a haven where he wouldn't be flea bait, always fearful of betrayal and guards, hungry every day, and miserable.

In the days following his failed rafting expedition, Stepan endlessly considered his options. He had picked up some of the language and often overheard the family talking admiringly about a revered man who lived not far away, Sheikh Hammud al-Aekleh. He was supposedly gentle and influential — and Stepan's only hope. So one February morning, Stepan rose earlier than usual to accompany his cows to the pasture. Then he left his animals behind and turned toward the area where the sheikh was said to live. With each stride, he grew more optimistic. In the distance, a throng of black tents appeared, meticulously woven from the fine hair of goats. Stepan had seen this type of home during his forced marches; they freckled the face of this golden desert and were sometimes called "houses of hair."

Amid all the tents, one stood out, a large structure divided into compartments. The sheikh's, most likely. Cautiously, Stepan crept closer so he could make out details. A man with a thin round face was leaning on a pillow atop a carpet, surrounded by his clan. He wore customary dress for winter, a *thob* flowing from shoulder to ankle, a wool coat, and a white headscarf ringed by a *faysaliyyeh* band.

Glancing up, the sheikh noticed Stepan in rags, hugging the periphery of the settlement. He beckoned to Stepan and asked what he wanted.

"I've heard your noble name for some time," Stepan stammered in Arabic. "And I have come to ask for whatever work you might give me."

"Where have you been staying all this time, and why do you want to quit your job?"

"Of course I work for my bread, but they don't give me enough and I'm always hungry. That's why I would like to quit my job and work for you."

"All right, all right. Sit down, my boy."

"Could I come tomorrow?"

"Why not today?" he asked.

"What I'm wearing was given to me by my employer. I want to go and return it; otherwise it would be stealing."

"Good for you, my boy. Then come tomorrow."

Stepan thanked him perhaps a thousand times before departing: *Shukrun, shukrun, shukrun.*

The next day, he left his life of fleas, pretending he was taking a short trip to Sabkha to buy clothes. In his own tattered rags, he embarked on the road that led there, but once out of view, he turned and headed in the opposite direction, toward the chieftain's residence. After half an hour, he arrived at the sheikh's tent.

"Sit down," the sheikh said. "What's your name?"

"Mustafa."

"No, no," he said, eyeing the wretched man before him. "I want to know your Armenian name."

"It was Misak," Stepan said, "but —"

"No, I won't permit you to change your name." He turned to the people around him, his expression strong yet quiet. "His name is Misak and you'll call him Misak."

Stepan couldn't believe it. After all this time, someone was accepting him for who he was, and he hadn't even given his real name. Having been formally registered in the camps, he was too afraid to give it, too afraid of capture again. The sheikh gestured to some of the others. A few people scurried away and then returned with a wooden tray covered with an enormous disk of wheat bread and *pekmez,* a type of syrup made from grapes. It was a feast large enough for many people, an entire boxcar of Armenians. Starving, but not forgetting his manners, Stepan waited.

"Misak, why don't you eat?" asked the sheikh.

"Aren't there others coming? Am I going to eat all this by myself?"

"It's all for you," replied the sheikh.

Stepan grabbed some bread, and the next few moments unfolded in a whirl of eating and dipping. He could not finish everything, despite the generosity. His stomach seemed to have shrunk. When was the last time he'd eaten this much? Not since Adabazar, not since his mother's table. His gratitude toward this man, this stranger, overwhelmed him, and he expressed it as best he could, "May God give you *bereket,* plenty, and good health."

Stepan carried the leftovers to the sheikh's wife in a separate section of the tent and set it down, like a servant. He thanked her too.

The sheikh brought Stepan garments. "Throw away your clothes." Stepan removed his rags, and he was shaved and styled the Arab way; his features were unearthed once again. He felt almost reborn, freed from the torment of insect scratches.

At lunchtime, he ate more of the *pekmez* and buttered bread, watching the activities around him. The sheikh's clan included the fifty-some families in the area, many of them his relatives. The sheikh's land was said to run from the barren cliffs down to the Euphrates, protecting thirty-five settlements. Given his proximity to the deportation route, the sheikh had taken in Armenians before, and he shared the story of the three other Armenians still living in his camp: two young women and a boy of about eight years old named Aris. All of them had been ravaged by the death march, like Stepan, malnourished and in tears until the sheikh nursed them back to health, caring for them like his own children.

Later that evening, the sheikh's brother, Ali, brought a lamb slaughtered in honor of two Arab guests who were about to arrive. Standing by while people were busy cooking, Stepan felt uneasy; he needed to prove his worth. "I want to do something to help."

"Misak, you sit next to the fire," the sheikh ordered. He was grilling a hulking side of meat, the juices welling up like tears, sizzling above the flames. He removed two chunks and passed them to Stepan, who was flabbergasted at the portion size. The servants diced the lamb, then dropped the pieces into a pot of boiling water while the women stretched out dough on a *saj,* a domed grill warmed by a fire underneath. When the wheat flatbread was done, the women tore it into fragments, set them on a large circular platter, and spread the cooked lamb, smothered in its juices, over it. Servants appeared with a basin of water and poured some onto the hands of the two guests visiting from another region. Stepan observed the sheikh and his companions sit down around the communal plate. They ate using their thumbs and fingers. The others waited their turn. The full ones eventually stepped aside for the next group, the order of eating based on importance.

The sheikh summoned him. "Misak! You sit and eat with them."

Astonished, Stepan felt great honor to eat second, especially given that the servants and workers made up the third and fourth groups in this hierarchy, dogs the fifth. Not long before, he had been grateful to suck on a watermelon rind. That night, he retired feeling safe and secure. In the morning, he volunteered to help again, but the sheikh once more instructed him to rest.

So Stepan returned to the Euphrates, his favorite place to think. How had he found this haven? This kind man? If he had ever doubted the existence of a higher power, he never would again. That evening, he realized his dire circumstances had changed. He could, he would, live. In the silence, he reflected on his ordeal. The rotting bodies. The screaming children. The good friends who had drawn their last breath. His sparing had not been a fluke but God's will. The waves of the river were beating down; he followed their crest and fall. Then he heard his name, "Misak."

Turning his head, he saw an Arab man approaching.

"Greetings, Misak," the stranger said as he extended his hand.

Who is he? How come he knows my name?

The man introduced himself as Hagop, a common Armenian name for Jacob. "I work for the Arabs in the village. Last evening, when I heard that another Armenian had come to work for the sheikh, I decided to come and meet you. I'm from a village near Bursa. You're very lucky to have such a noble sheikh to protect you. He's extremely influential and is very powerful; you have nothing to fear from the gendarmes. Stay here."

And Stepan did. The weeks passed, and under the sheikh's care and guidance, Stepan's health and spirits gradually rebounded. He and Hagop regularly met, and in time, he met the other Armenians under the sheikh's protection, including a young woman from the southeastern Anatolian village of Urfa, Lusi Markarian, whose company he enjoyed. As their relationship deepened, he held her close and kissed her tenderly. Neighbors inundated the sheikh with marriage offers for Lusi and the other Armenian girl, but he refused. Elsewhere, Armenian women were saved from the massacres by being forced into harems. With each refusal, the offers increased — fifty gold coins, higher still, sometimes one hundred. Still the sheikh would not marry them off, again impressing Stepan with his kindness. Stepan knew how many of the survivors were like him, just waiting to get home.

One day the sheikh asked him to clean his three rifles. As Stepan held the long handles and shined the frames, he could sense the sheikh watching him.

"Do you know how to use these rifles?" the sheikh asked.

"Yes, I do."

"Which one is best?"

"This is the best of them all," Stepan said, pointing out a new German one. The others, he explained, were older French and Turkish models.

"How do you know?" the sheikh asked.

"It's written on them. I read and found the maker's seal." With his fingertips, he traced the engravings on the weapon, revealing his own literacy.

The sheikh had been testing him, Stepan realized, to see if he really had the knowledge to differentiate one weapon from another. The sheikh wanted a consultant to advise him in his purchasing decisions, and that man was now Stepan.

As the weeks passed, Stepan steadily regained his strength. Only now did the chieftain allow him to work in the fields. He began with a starter herd of twenty donkeys along the riverbank but soon graduated to cows and then sheep, more responsibility. In the space of a month, his empire expanded to two hundred lambs.

As Stepan herded his flock one afternoon, a boy rushed toward him. It was Ali, the sheikh's son. "My father wants to see you," said the ten-year-old. Stepan followed him back to the sheikh, who held a firearm.

"Misak, they've brought me this rifle and want to sell it. The owner is asking six pounds, in gold. I offered four and he came down to five. I think that he'll even accept my initial offer. Is it worth it? Shall I buy it?"

Stepan made a long, close inspection of the gun and then rendered his verdict, "It's no good, don't buy it."

"The expert didn't like it," the sheikh told the seller, and he sent him away. His instinct had been right. Not long after, Stepan heard, someone else had bought it, but when he tried to shoot it, he learned it was a lemon, incapable of spitting out seeds. The sheikh heard this, too, and boasted of Stepan saving him the trouble. From then on, a procession of neighboring villagers began to consult Stepan about their weaponry, drawing on his reservoir of knowledge from his days as a courier.

One afternoon, Stepan noticed two gendarmes approaching on horseback. Stepan bolted to the tent's other chamber, and though he could hear the sheikh yelling his name, he did not turn back. The sheikh's wife approached his hiding place.

"Don't you hear him, Misak?" the sheikh's wife asked. "He's calling you. Why don't you go?"

"The guards are there and I am scared."

"Scared?" She laughed, and Stepan couldn't fathom why.

"Misak!" the sheikh yelled more loudly. Stepan emerged and cautiously made his way toward them.

"Were you hiding because of the gendarmes?" he asked. "Let them fear *you*."

"The sheikh is right," one gendarme said.

Drawing close, they addressed him in Ottoman Turkish. "You're very lucky that you enjoy the protection of somebody like him. We've been ordered to arrest Armenians wherever we meet one, but never here, never those with him. Don't ever move elsewhere."

Stepan could finally see the sheikh's true power. Even the ferocious gendarmes cowered before the chieftain like children. As the leader of his clan, the sheikh ruled his area like a king. Despite being officially under Ottoman rule, the locals of this region answered to him. He had the power to determine who lived, who died. In fact, for the longest time, this part of the desert was like the Wild West, filled with raids and skirmishes. Tribal custom had reigned. Even in the mid-1800s, when the Ottoman government built military outposts along the Euphrates to extend their domain and increase taxation, the Turkish officials paid Arab chieftains like the sheikh to retain control.

Nearly every day, these same gendarmes would visit the leader and pay their respects. Over drinks, they would all talk, the gendarmes and Stepan often switching to Ottoman Turkish. Eventually, during those long hours of bread and conversation, they stripped away the labels — Turk or Armenian, gendarme or deportee — and became just friends sharing a meal.

"Misak," the sheikh said one night. "Bring me some water to drink." Stepan studied the cups stored a few feet away. None seemed good enough. Turning to the sheikh's wife, he asked, "Which one should I use?"

She pointed to a porcelain cup. "Fill it and take it," she said.

Looking in, he was aghast at the ceramic's filthy state, and asked if it was okay to use.

"Fill it and take it," she repeated.

But he couldn't put water in such a dirty cup, especially for a man of such greatness. With fistfuls of earth, Stepan scrubbed the inside, removing the residue with natural abrasive. Then he rinsed it clean. The sheikh did not immediately drink. Instead, he stared into the cup, lost in thought. From that point onward, he requested that Stepan bring him his water, a task with the highest prestige for a servant.

As an attendant, Stepan was always hovering nearby, for the sheikh as well as visiting dignitaries. One night, while Stepan was sitting within view of a dinner party, the sheikh, in a joyous mood, motioned him over.

"Misak," he said. "We'll marry you to an Arab. What will you offer the girl?" The sheikh could not contain his laughter. He knew Misak did not have any material goods, only his borrowed clothes. The marriage custom demanded a sheep, and at that time, Stepan had been herding donkeys.

"I can give a donkey," offered Stepan as a joke. "If the girl is very beautiful, then two donkeys, even three, and up to five donkeys. But as the girls prefer sheep, I'll sell five of my donkeys and give her one sheep."

Howls of laughter. Stepan did not have even one beast to his name. "How many donkeys for this girl?" the sheikh asked, and pointed out a woman they knew. Hamming it up, Stepan carefully made his evaluation, calculating the worth of animal against the woman's attractiveness and then announcing the donkey count with a deadpan expression. Laughing, the sheikh pointed to another lady, and then another. The visitors' bellies were shaking, and their eyes tearing from laughter.

"Come now, Misak, tell me, do these donkeys that you're selling belong to you?" asked one guest.

"Of course," he replied. "It's even written in the registers of Sabkha as Misak's *mohr*." That was like a groom's dowry.

In the following days, romance thickened the spring air, and word drifted of Stepan's long-eared dowry. One after another, girls approached, their eyes dancing; they asked their worth and then broke into fits of giggles. In the evenings, Stepan turned his attention to Lusi. Often, he would visit her tent very late, the two talking in their native tongue into the dark hours of night, displaced souls finding some comfort in each other, far from home. All this — the interest from the Arab women, the late-night visits with Lusi — wouldn't have been accepted back home, where marriages were arranged, but what was home anymore?

That May, the Euphrates receded and the fertile riverbank widened, bequeathing a gift of land and room to plant vegetables. The sheikh wanted Stepan to oversee the harvest and had a shack built so that he would not burn under the sweltering sun. Stepan happily complied and continued to tend to the sheep and other livestock as well, doing whatever was needed. While working one day, Stepan heard girls singing and recognized his name embedded in the serenade.

"*Marhaba*, hello," they said.

"Misak," one asked, "would you marry me?"

"Yes," he replied.

"And me?" asked another.

"I'll marry you too."

The third one said, "Don't forget me."

"And you too," Stepan said.

The last — and most beautiful — girl put her hands to her heart as if to say, *What about me?*

"You," Stepan said. "Definitely."

"We don't want donkeys," the first three said.

"I will accept a donkey," declared the most beguiling.

"In that case," he replied, "I'll marry only you."

They were all joking. The prettiest one twirled toward him. "We have a request," she said. "Would you please get into the river and swim? We've heard that you're an excellent swimmer. They say you swim like a fish underwater."

Stepan undressed, laying down his long shirt on the shore. He swung his arms overhead and proudly dove into the water, his naked body disappearing under the iridescent splash. Eighty feet away, he resurfaced and inhaled. The girls looked impressed. "Bravo," they shouted and smiled as he stroked and kicked his way back toward shore, "Bravo." Modestly, he reached for his shirt on the bank while remaining in the waist-deep water. In that same moment, the sneaky siren stole off with it, leaving a trail of laughter in her wake. He fumed. After everything he had been through, his dignity was all he had left, especially after years of being naked out of desperation, not lust. His face burned. "Please," Stepan begged, to no avail. "Run after her and get your shirt," the girls teased. Stepan emerged from the water and chased her. As he caught up, she drew the shirt closer to her breasts, causing her friends to laugh harder. Confused and irate, he reached for the garment. The two wrestled, and she fell to the ground on top of the shirt. Someone pulled his leg, and then his arm, and he tumbled down beside her, conscious of his splayed legs and nudity, his body laid bare.

"I'll give you your shirt on the condition that you kiss me," she said.

"Is that all you want?" He stood up.

She did too, dropping the ransomed item to the ground. Pressing against him, against his wet hairy chest, she kissed him ardently; her mouth seeming to devour him. *"Ana ahebbak,"* she whispered. I love you.

"Enough. You achieved your *murad,* your wish," he said. "It's enough."

Ensconced in the sheikh's hamlet, Stepan had nearly forgotten that danger still stalked the nearby roads. One afternoon, the emboldened Stepan led his sheep farther away. As he climbed a hill, his eyes wandered to the road. He stiffened. Another *sevkiyat* was approaching; a flood of Armenians were advancing from the west, from Sabkha, their thin forms stepping toward Deir Zor. They had a few items on their shoulders. With the sheikh's protection, he was no longer afraid of the gendarmes, so he hurried down the slope in search of a familiar face. "Are there people from Adabazar among you? Or from Izmid, or Bardizag?" "No," they said. Looking at them, so desperate and sick, rendered him temporarily mute. "May God be with you," he finally managed, his eyes watery. *"Menak parov,* goodbye." As he was about to climb back up to his herd, three gendarmes galloped toward him.

"What did you sell?" they asked in Arabic.

"I didn't sell a thing," he replied in Ottoman Turkish. "I was only looking for people from my town."

"Ah, *giaour.* You're an Armenian? Get in the caravan. Quick!"

Stepan was stunned. This shouldn't be happening—he was supposed to be safe. Clearly, the guards didn't understand his status. "I'm a shepherd, Sheikh Hammud al-Aekleh's servant," he explained.

"We don't give a damn for your sheikh or *maykh* or whatever. Come on, filthy pig. Get in the caravan."

The pointed tip of a weapon dug into his back. The caravan was moving forward and Stepan with it. He scrutinized the guards. By now, he knew many guards from their visits to the sheikh. And yet when it mattered most, he didn't recognize any of them. He tried to threaten them. "Listen, the sheikh's sheep are left unsupervised. There'll be a lot of trouble for you. The sheikh is powerful."

They didn't respond. He glanced around; no one of use was in sight.

Why wasn't anyone listening to him? He panicked. Ten minutes passed, ten minutes of repeating the mantra "I'm Sheikh Hammud al-Aekleh's servant" to a gendarme. The man ignored him. Stepan tried another, "I'm Sheikh Hammud al-Aekleh's servant."

That gendarme responded by slamming a rifle down on him.

Stepan knew this road to Deir Zor all too well. Twice before, along this route, he had escaped the convoy by pretending to urinate, using darkness as a cloak. But in daylight, he couldn't flee, the arid plain too bright. He focused on the guards again, still puzzled why they didn't know the sheikh's name. In the next hamlet, the march slowed, and two gendarmes joined the caravan, carrying food. Stepan made eye contact with one, and his joy nearly leaped from his throat.

"Misak! Is that you? What are you doing here among these people?"

"Your friends caught me and put me here."

"Didn't you tell them who you work for?"

"I did, but they don't care about the sheikh or anybody else."

"My God, we're dead! What are we going to tell the sheikh . . ." He stormed off, huddled with the gendarmes, and yelled and cursed so loud Stepan could hear him from where he stood. A short time later, he reappeared. "Return to your sheep," he said in front of the other prisoners and guards, "and please say nothing to the sheikh."

"Why shouldn't I tell him?" Stepan asked. "What happened to the sheep? I'm responsible."

"We'll come back in a few days and apologize to the sheikh," the guard said.

Stepan turned and clambered up the hill. The guards weren't following him, as far as he could tell. He searched for his herd, his eyes darting around the dry brush and sand, luckily finding them grazing in the same spot as if nothing had happened and the past few hours had been imagined. Shaken, Stepan relayed everything to the sheikh that evening. The sheikh looked confused, "Did you mention my name?"

"I did, but it made no difference."

"Will you recognize them when you see them again?"

"Yes, I will."

"Good. Then you'll see."

A few days later, the same guards approached on their horses, kicking up veils of dust. Still some six hundred feet away, the gendarmes halted and dismounted. They fell to their knees and crawled to the sheikh, who stood still as a statue, not uttering a word. Once in front of him, they prostrated themselves on the ground, as if in prayer, and pressed their lips to their host's feet. Sheikh Hammud al-Aekleh's expression remained unchanged, his eyes fixed on the horizon. From the sidelines, Stepan watched.

"Misak, come and show me who forced you to join the caravan," the sheikh said. "Show me the one who heard my name and didn't care."

"These are the ones," Stepan said, pointing out the men who had apprehended him.

"Tell me, what do you want me to do with them?" the sheikh asked. "You decide their punishment."

Terrified, the guards turned to Stepan, this minuscule man in shepherd's clothing. Begging for mercy and forgiveness in Turkish, their voices trembled, their faces marked by fear. As Stepan listened, the fearsome guards seemed to shrink in size, defenseless now, like he'd been. His heart broke. He didn't want more death; he had seen enough of that for one lifetime, for many lifetimes.

"I forgive you."

The men did not say anything; the silence was filled by their labored breaths. Only the sheikh spoke, "You owe him for your going unpunished."

Two Hammuds

2007

IN THE CAR, midway across another bridge, I looked down at the now-dammed Euphrates flowing placidly, the banks bordered by lush green meadows. So different from my grandfather's time, I thought, when the river nearly bucked him off his raft of corn stalks, preventing him from getting to the other side.

Now we reached the left bank easily and entered the city of Raqqa, the streets widening, flanked by sunbaked homes and scattered ruins. We wound around a traffic circle, and plaza, festooned with the colors of the Syrian flag, and crowned with a tall gold statue: President Bashar al-Assad's father, Hafez, who ruled the country for nearly three decades. He greeted visitors like King Kong, sturdy and strong, seemingly unshakable. Almost everywhere in Syria an Assad stared back at you, usually photos of the thin-mustached son Bashar, his blue eyes reminding that he, or at least someone, was always watching.

Since arriving in the country, I had felt the president's authoritarian reign everywhere, as well as a quiet discontent by the people. Some whispered when they spoke, or checked both ways to make sure no one was around. At the moment, though, I wasn't thinking about the state of affairs here; I was singularly focused on Sheikh Hammud al-Aekleh. He had lived in this desert, beside this very river. Amid the bluffs and brown rock, I'd looked for signs of him, wondered where he used to eat and sleep, wondered if his family was still here.

As we drove down the streets, the sidewalks bustled with pedestrians. Spires of churches towered alongside the minarets of the mosques as people of all faiths lived peacefully together. The car slowed, and my driver, Antranig, parked in front of a beige two-story house with twisting columns and decorative stone, girded behind brown bars. Levon had directed us here after telling us about a Bedouin chieftain who commanded a clan of thousands and might be able to assist me in finding Sheikh al-Aekleh's clan.

At the front gate, which was gilded with palm trees, I waited. The meeting had been arranged only the day before, and I didn't know the local customs, how he'd react to my request.

An Indonesian servant opened the door and escorted us to a large rectangular room, a formal receiving area. Carpeted with Oriental rugs, there were no couches, only plush cushions along the wall, with movable armrests. Although I hadn't been in a traditional Arab home like this before, I knew enough to lower myself to the floor. A few minutes later, the sheikh entered the room, wearing a long robe. At once, Antranig and Levon rose to their feet, as if the sheikh were a judge taking the bench, and I hastily followed, nearly tripping on my skirt. With a thick mustache and round glasses, the sheikh, Fayez al-Ghubein, uttered a greeting in Arabic, his expression stern and intimidating, one that commanded respect for his seventy-some years. He wore a white *kaffiyeh* on his head, crowned with a black *agaal*, a black ring, his salt-and-pepper hair cropped underneath. The white *thob* draped over his body, loose and flowing like a cape, as he strode across the room and sat opposite us.

"Why have you come?" he asked as Levon translated. Naturally, he was curious, because few foreigners visited this region. I told him about my grandfather and his caravan's trek through here with neither food nor water and how I'd needed to see this land for myself. As I rambled, I could see his features soften. "My family took in Armenians. Around twenty," he explained. His clan was part of the Anizah tribe, which had roots in the same region as the Saudi royal family. He'd grown up in the

shadow of the Armenian death march, hearing the stories from his elders about his clan's compassion, a lifesaving act that wasn't forgotten.
Two years earlier, he explained, the Armenian government had honored his clan in Yerevan, the capital, and he'd traveled there for the
ceremony. Now he retrieved a framed photograph in which he stood
beside chieftains from all over the region, all lauded for their forefathers' bravery.

I felt so heartened to think about all these other heroes who'd saved
Armenians, just like Sheikh Hammud al-Aekleh. Seeming to trust us,
Sheikh al-Ghubein invited us to his private chambers. His beautiful
wife greeted us and ushered us onto the red brocade divan. The servant
brought in an afternoon meal of marmalade, dates, flatbread, and sweet
hot tea, spreading out the food on a small table in front of us.

"Where are you staying?" Sheikh al-Ghubein asked. Since we hadn't
made any arrangements, he insisted we remain with him. I tried to refuse, not wanting to overstay my welcome, but when Sheikh al-Ghubein spoke, it felt like a decree. As everyone picked up in Arabic again,
I nibbled on some succulent dates. Spitting out seed after seed, I grew
increasingly anxious about my quest. Hours had passed, and we hadn't
even broached the topic of the *other* sheikh, Hammud al-Aekleh. I
waited for a break in the conversation, and when more flatbread and
tea arrived, I cut in.

"During the massacres, my grandfather was about to die but was
saved by a sheikh in the area. He took him in when he was hungry and
thirsty. I want to find his family and thank them. If there is any way you
could help me, I would be so grateful." He sat there listening, his face
serious. I wasn't sure if I was speaking out of turn, especially since I was
a woman, and especially given his position in this region; perhaps I'd already imposed on his generosity. I pressed ahead anyway and asked if
he knew of Sheikh Hammud al-Aekleh.

"What village?" he asked as Levon translated.

"I don't know," I replied. "But I believe it was around Deir Zor. My
grandfather didn't write the name."

Just the day before, I wouldn't have been able to answer this question. For months, I'd been studying my grandfather's notebooks, re-

reading key passages detailing direction and geography. But the night before, in preparation for this visit, I'd followed his twists and turns on a map and thought I had identified where my grandfather might have ended up.

Sheikh al-Ghubein rose and dialed a number. Not long after, another man, tall and handsome in an olive-green *thob*, appeared at the door of the private chambers. He began working the phones, canvassing the region, using both the landline and a cell, one to each ear, as he called the various clans. As soon as he ended one conversation, the phone would ring again, and he sometimes spoke on both lines at the same time. I was hoping for some indication of his progress, but his expression betrayed nothing. My heart pounded, thumping in my ears. Finally, he lowered one receiver, and said, "There were two Hammud al-Aeklehs."

"What?" I shrieked as if I'd just won a sweepstakes.

"One in Sabkha, and another in a small village. Do you have any more information?"

"My grandfather said Hammud al-Aekleh was very powerful," I said. "His brother's name was Ali. His son's too. Ali would have been a boy during World War One."

He continued in Arabic with his source and then hung up. "In Sabkha, Hammud al-Aekleh didn't have a brother named Ali." He dialed the other family. I was nervous and exhilarated. What were the chances? It was a long shot, I knew. But now, I watched his face grow with animation.

"Have you found them?" I asked.

"Yes," he said and smiled. This Sheikh al-Aekleh had both a brother and a son named Ali. Electrified, I couldn't believe it. These little clues my grandfather left, crumbs from the caravan trail, had led me there across the decades. The village was situated just as I'd thought after studying his journals and escape route. I felt as if I were emerging out of an *ajaaj*, out of the sandstorm of self-doubt that had obscured my way. The path forward seemed clear now.

"Can I see them?" I asked. Goose bumps prickled my arm.

"Tomorrow? Is that okay?"

Crossroads

1917

THE DAYS THAT HE DROVE HIS BLEATING SHEEP TO GRAZE, Stepan was on high alert, scanning the riverbank for livestock thieves hiding in the tangles of vegetation at the water's edge. The Bedouin marauders had struck once, and now Stepan carried a rifle at his side, his protective instincts roused. Somehow, in this bone-dry desert, he'd found refuge—and it wasn't a mirage. Holding his weapon, he felt proud, like a real Arab, like one of the sheikh's sons. And he wasn't afraid.

Yet he couldn't shake the loneliness. He hadn't seen his family since the fall of 1915. Were they alive? As he pondered this time and again that summer, more Turkish military boats cruised southeast down the Euphrates, loaded with ammunition and weapons, and troops advanced along its banks. The flotilla was supplying firepower somewhere eastward. The soldiers rode with the cargo to Deir Zor and then marched back to their starting point on foot. At least, that's what he heard from a passerby. But where were the arms going from there? That March, Baghdad had been lost to the British.

Stepan didn't know that the Turks had launched an offensive to reclaim the ancient city, an operation they dubbed Lightning, or Yildïrïm. Ottoman soldiers were concentrating in the region, readying for an attack on British positions. With troops already stretched as thin as phyllo dough, the Turks were dangerously diverting resources from other critical zones, such as Gaza, the gateway to Palestine, where they expected yet another British assault.

Many soldiers passed through the sheikh's lands in those warm months. Stepan was still cautious despite the protection of his benefactor, but there was one sergeant, Hasan Chavush, he looked forward to seeing and thought was sincere. He reported to a base in Birejik, more than a hundred miles away, and commanded a handful of men. Somehow, between the numerous trips to the desert over many weeks, Hasan and Stepan, an Ottoman Turk and an Ottoman Armenian, a Muslim and a Christian, became friends, as in the period before the war.

One afternoon, a hapless man appeared in front of Stepan. He was alone and clearly didn't belong to the troops of passing soldiers. The stranger wore very little clothing, leaving most of his body exposed. Stepan called to him in Arabic. He didn't seem to understand.

"Please don't kill me," the man begged in Ottoman Turkish, his fear rising as his eyes locked on Stepan's weapon. "Please don't kill me!"

Stepan answered him in fluent Ottoman Turkish, not Arabic. In an instant, the man knew that Stepan wasn't the armed local he appeared to be, that he was yet another displaced Armenian.

"Ah, *arkadash*, comrade, God has sent you. Glory to the Lord," he cried. He was a Turkish deserter, Stepan learned, fleeing the military. After being robbed several times, he had lost his way.

Stepan described the safe roads back to Aleppo, and not long after, one more deserter crossed his path. This Turkish soldier explained he'd been on the Russian front. The battles in the Caucasus had been intense, but all that was swiftly changing with the revolution sweeping Russia. Already, the Russians, tired of war, were beginning to withdraw from eastern Anatolia — and from the conflict as a whole.

The soldier needed directions to Aleppo. Intimately familiar with the roads, Stepan told him the way, and the two parted. Sometimes, the local Arabs brought disoriented Armenians to Stepan. Terrified and shaken, they seemed like lost lambs that had escaped the slaughter. Greeting them in his mother tongue, Stepan sheltered them overnight and then pointed them home. From them and other passersby, he'd learned the near impossible: the massacres of Armenians had ended. Going home was possible. As they set out, he felt pangs of jealousy. He wanted to follow.

One day, the shepherds arrived not with Armenians but with news — the suspected livestock thieves were back. The roaming Anizah blew about the region through the year, migrating from Central Arabia to this area en route to Aleppo. The Bedouin tribe always came before spring, as the Euphrates steadily rose from snowmelt, and then they circled back for the cold months. The seasonal migrations, however, strained relations between the two groups, the nomads and the settled.

Now the nomads camped not far from the sheikh's village. This time, they came from the Baghdad region, their rotation perhaps altered because of the conflict. With the tribe's reappearance, the sheikh drew a line in the sand: He wouldn't lose any more livestock. Dispatching messengers to his thirty-five settlements, he rallied his men. By the next day, four hundred of the sheikh's Arabs galloped into his encampment on horseback, brandishing their weapons, ready for battle. Before them appeared the sheikh, his solid body protected in armor, as he prepped his warriors. He mounted his horse and gallantly led his army toward the Anizah's camp, all his men following except for Stepan, still too weak for combat.

With the fighting under way, only women and children milled about the village. It was an opportunity, Stepan knew, to leave. He needed to know if his family was alive. He needed to tell them that he was alive. He felt horrible, knowing that the sheikh counted on him, not only for the sheep and the rifle consultation, but for other important matters, such as medical advice, like the time when the sheikh's son Ali mysteriously fell ill. As soon as Stepan heard, he had rushed to the boy's bedside and administered a concoction of warm milk, followed by cold milk, followed by *madzun*, or yogurt, curing the boy of his ailment. The sheikh had called on him again when his own brother, also named Ali, became incapacitated after a gluttonous dinner party.

When Stepan saw Ali Agha lying in bed thrashing his arms and legs, he felt anxious; he hadn't graduated from third grade, let alone medical school. Touching Ali Agha, he detected a fever. After talking with Ali's wife, he learned that the night before, a lamb had been butchered in honor of guests and that Ali had eaten voraciously. With that new intelligence, Stepan felt a burst of confidence and served Ali a glass of milk, followed by *tan*, liquefied yogurt. By the next morning, Ali had cleared his bowels and felt better, and Stepan thanked God for the guidance. His reputation as a healer was cemented, and the sheikh was deeply grateful. Stepan had built a new life here along the river, he knew, a good life, but he was at a crossroads now, unsure which way to turn.

That evening, as darkness hushed the campsites, a group of soldiers arrived. Not the returning fighters, but Hasan Chavush and his men. Having delivered their ammunition to Deir Zor on boats, they were staying overnight before pushing on to their base in Anatolia. Stepan saw a plan begin to take shape. He motioned Hasan over to talk, and the two strolled away from the tents.

"Hasan Chavush, I can be open with you and have a request. Would you take me with you, now that the sheikh is away?" Stepan couldn't bear to face the man who had saved his life and tell him he was leaving.

"May that be your only problem," Hasan replied. "There's nothing easier. We'll be heading out early tomorrow morning. Go to bed now. I'll wake you up before we depart." It seemed like the night's stars had aligned to form an arrow home and all he had to do was follow it. There wasn't much time. Stepan prepared for his last night in the village that had become his second home. This was one of the most agonizing decisions Stepan would ever make, walking away from the steady flow of flatbread, the soft bedding under his body as he slept, the great water source that had given him life after his desert crossing. He'd be abandoning all this for the unknown, for a gamble on a family that might be dead. He'd miss his daily greetings with the sheikh and his clan, the respect he felt from the great man himself, not bestowed on him since his courier days. He also knew that this was a rare opportunity. If he didn't go now, he never would.

His thoughts then turned to Lusi, the enchanting Armenian girl. Could he disappear and not tell her? In the darkness, he followed the well-worn path to her tent, long after the others had fallen asleep. Lusi often stayed up late, and that night was no different. He found her stretched out on the ground, wide awake. Crouching beside her, Stepan broke the news. "Lusi, I have a secret to tell you, but please, don't tell anybody."

With these words, her face lit up. *Oh no*, Stepan thought, *she is probably expecting a marriage proposal.* Understandably, since the two had grown close; perhaps he'd led her on.

"Tell me," she beseeched and stared intently into his eyes.

"You know, Lusi, we live here comfortable and safe, but I have family elsewhere and they don't know that I'm alive. And I don't know if they're alive or not."

She didn't say anything. She was beaming, transfixed by every word.

"With the sheikh absent, and Hasan Chavush here, I've decided to leave with him."

Her expression turned to anguish, and tears rushed to her eyes. "I depend on you and now you're leaving me alone." She was right, and he was doing the same thing to the sheikh.

More tears fell, and Stepan wept as he said goodbye to her, to this sanctuary, to his life there. "I haven't got much time. I must leave . . ."

Her wails grew louder, and Stepan worried that she would wake the others. She hugged him, kissed him, and he returned her embrace. He could feel the warmth of his own tears now storming down both cheeks. Nothing seemed to stem the tide of her pain. Although each had been deported here separately, they had found comfort in each other. But precious time was elapsing. Hasan would be leaving soon, and Stepan needed some rest for the long journey ahead. Lusi wrapped her arms around him so tightly he felt immobile, imprisoned by sadness.

Summoning all his physical and emotional strength, he pried her arms away. He had to find his family, even if this meant betraying the sheikh, Lusi, all the people he'd come to care about.

"Goodbye," he said one final time. "Please, don't say anything to anybody about my departure. May God be with you."

For the rest of the night, he listened to her muffled sobs nearby. Quietly, he mouthed the words he wished he'd said. Then, just as swiftly, he changed his mind — he shouldn't have told her anything. Steeped in regret, he drifted off to sleep, and the next thing he knew, Hasan was shaking him. "Misak, hurry, get up, we're leaving." He could barely see in front of him; the sun hadn't yet risen. Springing up, Stepan grabbed his sack, and he hurried toward the others in the darkness. He set off, never forgetting the kindness he had found in the desert, in the land that should have been his burial ground.

Hours of walking, and the first glimmers of dawn brushed the sky. "Take off your *mahrama* and *agaal*," Hasan Chavush instructed Stepan. Away from the sheikh, Stepan knew he had to be careful, so he slipped off his Arab headdress and put on the used military cap and coat Hasan handed him. *Blending in,* he thought as they marched west, in step, eventually encountering a pair of gendarmes. Stepan tensed, but the danger passed.

The towns of Ma'din and Sabkha came first, then a series of indistinguishable hamlets before they reached Raqqa, where he'd rotted in that roofless cell and, once, run into Khoren. Hamam was next. On the fringes of it, the men camped, continuing at dawn until crossing out of Deir Zor region. Finally, he'd emerged out of that black hole. Innumerable friends had not, forever lost. It was time to change again. He swapped his fatigues for his Arab robe and headdress, so threadbare after months of continuous wear they could hardly be called clothes. *The most dangerous areas are behind me,* he thought as he saw more ragtag campsites dotting the riverbank, the signs of Armenian survivors.

"Hasan Chavush," he said, "I'll go and look for some Armenians among these tents."

"Don't be late; come back soon."

A few young boys were fetching water from the Euphrates. Stepan approached and listened, their dialect clearly from Bardizag, a town near his own.

"Is there anybody from Adabazar here?" he asked. Startled, the boys stared at Stepan in his ragged Arab garb.

"Ah! The Arab speaks Armenian," they finally said. "Yes, there is — the *yemeniji*, the slipper maker, Bedros Tozalakh."

Of course Stepan knew him. "Show me his tent," he instructed, "but from a distance."

The boys pointed out a tent whose walls were made of a thick weave.

He recognized the rag carpet from his region. He entered the tent and discovered a man and woman.

"*Parev*," he said. The couple, seated on the ground, appeared thoroughly bewildered at his hello.

"Don't you recognize me?" he asked.

The husband and wife turned to each other, perplexed by the appearance of an Arab speaking Armenian.

"I am Stepan Miskjian."

"What?" The two jumped up. "Is this Stepan?"

"Before everything else, I'd like to know if you have any news about my family. Are they alive? Where are they?"

"Yes, they're alive."

They're alive. It wasn't a question anymore, it was a statement: *They're alive.* This was the news he'd been waiting for, the hope that had never dimmed. Pure joy was now looping through him, traveling to fingertips, to his toes, to his light head. He tried to picture them. Alive, yes, but what were their conditions? Were they gaunt like him? The husband told him more now, that they were healthy and still living in Bolvadin, just miles from Chai station, where he'd seen them. At long last, he knew their fate and could share with the man a bit of his travails, including his rescue by the sheikh, the man to whom he owed his life. He would have kept going, but Hasan appeared in the doorway.

"*Menak parov,*" Stepan said to his compatriots. Goodbye.

Westward, the men plodded. Hasan couldn't help but notice Stepan's giddiness. "Misak, you're looking very happy."

"Yes, I am happy, because I heard that all of my family is alive."

"My friend," he said, "for this cruel destruction perpetrated by our people against the Armenians, even our grandchildren won't be able to compensate. The future of our nation is dark; that's how I see it. Sooner or later we must bear the punishment."

"The real criminals are your leaders," Stepan said.

"The entire Turkish nation is guilty before the Almighty and will be punished," Hasan said. "There's no salvation."

Stepan didn't say anything in return; he just kept marching. That night the men camped in a small village and then pressed on at first light to Meskene. At the edge of the Euphrates, they rested near a large market bustling with shoppers.

Stepan set off into the tangle of traders, searching for more natives of Adabazar. In a shanty butcher shop, he recognized an old friend from the labor battalion, Krikor. Having feared the worst for his work mates, he was astounded. With another *parev,* he startled Krikor and the crowd with his Armenian greeting. Gathering around him, they ogled him in his *mahrama* and *agaal* as if he were an exhibit at the zoo, murmuring, "An Arab-looking Armenian."

"I am Armenian," he said out loud, the words ringing through him. "My name is Stepan Miskjian." Gone was the name of Misak, Mustafa, all his aliases, all but his own. His pride and relief couldn't be contained.

"Oh, Stepan!" Krikor exclaimed. "What's this?"

A young man in the crowd who recognized Stepan's name dashed off and returned with a message, "You'll come to our house. My brother-in-law wants to see you."

Onnig Daylerian. *Aman,* his old friend from Adabazar. Stepan followed the young man to a small hut where Onnig, his brother-in-law's best friend, stood outside. Onnig began to cry, the tears streaking his cheeks. "I never thought I'd see you like this."

"What are you talking about, *akhbar,* these are my best clothes. You should have seen me in the desert."

They entered the shack, and Onnig told him all about Aghavni's letters. How Stepan's sister had written him constantly, asking about Stepan. "I always wrote back, saying that 'Stepan has gone far down south, and despite my inquiries, I haven't been able to get any news of him.' In her last letter she asked why I didn't write anything specific; 'If he's dead, say it.' I had to tell her that there was no news about those who had gone south, that there were no survivors. Now write a quick note to her and I'll send it."

At long last, Stepan had his family's current address and could write the letter he had been dreaming of for so long: *I'm alive!* Another Adabazartsi stopped by, Haji Hagop Semizian, one of the wealthiest merchants of his town. Before the exile, he had run a wholesale-grocery business and had been among the first expelled from Adabazar. When

Stepan finished the note, Haji Hagop took it into his hands and promised to send it to his family, who resided close to Stepan's. It was far from where Stepan was, but he promised himself he'd find his way to them before long.

Hasan Chavush returned. As they parted, Haji Hagop offered Stepan some tobacco and all of the change in his pocket, twenty-eight silver ghurush. Stepan refused, but Haji Hagop insisted. Touched, Stepan knew that was a lot for Haji Hagop; without his business, he was like everyone else now. Neither man knew that one day Stepan would marry this man's eldest daughter, Arshaluys, which means "dawn," a new beginning. For now, he had to go.

On the road, puffs of cigarette smoke rose above the men as they marched, the plumes of Stepan's celebratory tobacco. With Haji Hagop's cash, he bought food, and he shared that too. He was in a giving mood. *My family is alive!*

In two days, the men arrived in Hasan's village to rest. Afterward, Hasan had to continue on to his base, and he arranged for the village elder to host Stepan. That night and the next few, the Adabazartsi slept on the floor of a mill under a sackcloth, under the man's protection. However, it wasn't safe there, he discovered, since gendarmes were scouring the area for deserters. After a close call, he fled from there, too, and arrived in the old port of Birejik just as the sun rose over the Euphrates, streaking it in blue, such a contrast against the surrounding mountains. The town was awakening; the market too. Down the line of shopkeepers, he stumbled into some Armenians and inquired about his townspeople.

"Not here, but in Jarablus," one replied. "You have to cross the Euphrates. The road begins after the bridge. To cross the bridge you have to pay ten *paras*."

Stepan didn't have ten *paras*. However, he wasn't deterred. He would swim across, he decided. With the morning chill still tingling his skin, he waited on the roadside for the air to warm. Another man soon appeared, a man he knew was Armenian. "Excuse me, my friend," Stepan said. "Do you know anybody from Adabazar, Izmid, or Bardizag here?"

The passerby looked quizzically at him and then said, "Stepan?" Stepan just stared back blankly. He didn't know who this man was, but then it hit him: Vartan Chullerian, from Bardizag. Stepan couldn't believe it. The two had served together in the labor battalion in Izmid. First Krikor, now Vartan. He, too, must have escaped.

"Come, let me take you to Harutiun Efendi Atanasian," he said. Why was he bringing up Harutiun? Stepan had last seen his friend near that hill, the night he had clambered out on his hands and knees and run for hours straight.

"No, my friend," Stepan corrected. "Atanasian died in the massacres. I escaped but he refused to, despite his wife's pleadings."

"No. Atanasian also escaped the massacres; he's here. He's the German director's translator at the shipyard."

Stepan stood there, stunned. "So he fled too?" he said aloud as Vartan led him to Harutiun. Stepan's mind churned: Had Harutiun made it to Raqqa? Perhaps he had been swept up in the mobilization when Operation Lightning began. That was when twenty-five hundred Armenian deportees were recruited to work in the boatyard in Raqqa. In the distance now, Stepan recognized his friend, the high forehead, the glasses, the intense dark eyes. Harutiun moving, walking, breathing — alive. Wanting to surprise him, he quieted Vartan until the three stood facing one another.

"*Parev,* Harutiun Efendi," Stepan said.

Harutiun studied Stepan's sunken face, obscured by a nest of curly, knotted hair. He seemed to reach for words only to have them slip away. "Stepan, is that you?" he finally managed.

"That last night," Harutiun said, "when you proposed that we escape together and I refused, in spite of my wife's urgings, I had altogether accepted the idea of dying. But after you left, my wife began nagging me, 'Stepan left, you also should leave. You get away and be free; we'll find each other after.'"

A few hours after Stepan, Harutiun followed, slipping through the Chechen line, still not quite sure how he'd managed it. Wandering the desert for six days, he similarly encountered others, only two of them making it to Raqqa. "It was a miracle that I survived," he said. "I only wish my wife were with us; she might have been saved too. Do you think I'll find her?"

The question hung in the air. Deep down, they both must have known the answer. Changing the subject, Stepan recounted his own odyssey, the one that mirrored so many other Armenians' sagas of courage and resilience summoned from the deepest well of the soul. "I was also saved by a miracle," he said.

Seeing that Stepan was hungry, Harutiun fed his longtime friend, clothed him, and treated him to a shave and a haircut. As the hair tumbled to the floor, Stepan began to feel like himself again, like a human being. Harutiun suggested he work for the military, as counterintuitive as that seemed. "It's the safest way to live," he explained. Times had indeed changed. The Turks needed all the manpower they could get — and Stepan needed to eat.

At the nearby quayside, Stepan gazed at the enormous operation. An army of a thousand laborers toiled over boats in various stages of construction. The induction process in the director's office was swift.

"Ironmonger or carpenter?" the *efendi* asked Stepan.

"Neither."

"Then we'll place you in the *kalafat,* caulking, division."

The vessels were not that big, Stepan soon discovered, about twenty feet in length, with six-foot beams. Stepan stuffed rags in the hulls' seams, nailed them down, then coated them in tar, readying the boats for the voyage down the Euphrates to Deir Zor; from there, the cargo would be carried overland to the battles between the Turks and the British.

With a mailing address now, Stepan wrote to his family again as the late summer turned to fall, notifying them of his whereabouts. After an agonizing wait, he finally received the reply he'd been hoping for, in that familiar penmanship. His first letter, his siblings explained, had come as a complete shock, since they had been told he'd died and had been mourning his passing. How could the dead write? Not daring to fully believe it, they hadn't told their mother of his letter. Only after receiving the second letter did they really accept its truth.

Comforted by his correspondence, he threw himself into work. Production at the boatyard shadowed the rise and fall of war — sometimes forty craft a day were launched, other times, the hammers nearly halted. At one point, the caulkers were reassigned to a spot near Alexandretta, a port on the eastern Mediterranean coast. Departing immediately, Stepan and two hundred others first marched for five hours to Jarablus and then traveled to Qatma by rail. Stepping down onto the soil, now an Armenian burial ground, he recited a *hokvots*, a prayer for the dead, "Sleep in peace, dear compatriots. You have been sacrificed and deprived of freedom, but we will see freedom and never forget you."

Three hours later, Stepan and his fellow laborers arrived at the thoroughfare they were to repair, its surface cratered like the moon. As Stepan sweated alongside a mostly Turkish company, dumping dirt and stone into the holes, his only set of clothes ripped more. In the mornings, German soldiers arrived at the worksite by freight car, and nights ended with them leaving the same way. Communication between the groups was impossible; no Turk could speak German, and vice versa. Exasperated, the sergeant major lined up Stepan's unit during lunch one afternoon and asked who could speak French to the bilingual German officer. No one stepped forward. The *emanetji* considered raising his hand, but his knowledge of the language was rudimentary. Weaving among the erect soldiers, Muharrem Efendi, as he was called, grew increasingly infuriated, "Oh God! What I want is not something extraordinary!" With the bar seemingly lowered, Stepan volunteered — and the commander howled his fury for his not speaking up earlier. "Donkey!" he shrieked, and he grabbed Stepan by the collar and dragged him to the German. The sergeant major wanted the Germans to know that they shouldn't come on Fridays; that was a day of prayers for Muslim Turks.

Stepan stammered the message out, and the German seemed to understand. Still, Stepan worried until Friday, when, by grace of God, the Germans didn't appear. Tapped to be the German's interpreter, Stepan felt a surge of pride and soon realized that his shaky French was far superior to the German's. The job entailed accompanying the German officer back and forth to the Qatma base. During one of these trips, Stepan encountered a German captain who could not stop staring at Stepan — his ripped bottoms, his exposed buttocks, his sunken cheeks. The captain said something to a nearby soldier, and the next thing the *emanetji* knew, he was eating bowl upon bowl of meat broth and bread. He hadn't seen this much food since he'd left home; it was even more than he'd had with the sheikh. "God, how lucky can I be!" Stepan told himself as he rejoined his unit, feeling renewed.

While Stepan was busy translating Ottoman Turkish into French, the course of the war shifted. By year's end, the Americans had joined the Allies and were fighting on the western front. And with the Russian Revolution, the country's new leaders began deliberations to withdraw from the conflict and drafted a peace treaty with the Germans. In the Ottoman Empire, Palestine was also a battle zone. Gaza fell to the British after the third attack, and in December, Jerusalem came under British control.

To clear the injured from the war zone, the Turkish soldiers returned home on special locomotives.

One such train traveled west along the tracks after departing from Ras al-Ayn, the slaughterhouse. The doctor aboard the train, Garabed Kabadayan, treated these patients, who seesawed between life and death. Though he was an Armenian from Adabazar, he'd avoided deportation because of his medical skills. He'd also managed to keep his brother Dikran close by, as an enlisted soldier. Jerablus was their next stop.

Stepan's group was also on the move. To reach their destination of Birejik, they took the train to Jerablus, a transfer station; they arrived late in the afternoon and decided to sleep there before setting out again on foot. As Stepan sat on the platform the next morning, a train full of wounded soldiers pulled in. Some climbed off the train, injured but mobile enough to stretch their legs. The others remained inside, possibly too weak to walk.

As the men milled about the station, Stepan spotted Dikran Kabadayan. Ecstatic, he rushed up to greet him. Stunned and saddened by

Stepan's condition, Dikran had an idea and excused himself. "I'll go get my brother. He's in charge of these soldiers."

He returned with his brother Garabed.

"What's this, look at you!" the doctor said.

"*Eh,* I'm a soldier," Stepan said.

"Where's your family?"

"In Afyon Karahisar *vilayet.*" That was Bolvadin's province.

"If you'd like, I can take you all the way there with my troop," the doctor said.

The good physician couldn't bring an Armenian onto the train, so he handed Stepan old military clothes and yet another identity: Mehmed İbrahimoghlu, a Turkish soldier who had died in the war. With these papers, Stepan would get home. With the help of the Christian doctor and all the Muslim men who had come before — Hasan Chavush, the sacrificing gendarme at the river's edge, the sheikh — he could survive the final year of the war.

The Feast

2007

OUR CAR WAS KICKING UP DUST, rocking from side to side. The ruts on this dirt road were too deep for our tires. Only minutes earlier, we'd turned off the main highway through a canopy of landscaped trees, and we were already deep in the tiny village around Deir Zor, the onetime home of Sheikh Hammud al-Aekleh. Coughing, I closed my window as a whirl of concrete homes, cornfields, and flocks of sheep sped past outside. Nervous about my meeting, I smoothed my hair and began to fret: *Would his family like me? Would I make it past tea?* I expected a great-great-grandson, a wife, a child or two, a reception that would last about an hour.

As we rounded a bend of mooing cows, I spotted a crowd of some three hundred people. A religious event, I thought, related to the mosque nearby. Antranig continued driving and shut off the engine in front of a modest, single-story house. Beside us, the distinguished Bedouin from Sheikh Fayez al-Ghubein's home parked his blue car. He'd arranged this meeting and we'd followed him here, along the Euphrates. His car was always easy to spot because of the picture of President Bashar al-Assad on its rear windshield. As I swung my door open, dozens of men and women and children ran toward me. Everyone was talking at once, the highs and lows of Arabic overlapping and rushed. I felt big kisses on my cheeks, hands on my shoulders, the warmth of their touch. Slowly, I stepped forward into more outstretched arms.

"This is the family of Sheikh Hammud al-Aekleh," Levon pronounced grandly.

"All of them?" I asked.

"Yes," said the usually unflappable Levon, seemingly as awed by the moment as me. I looked at the men, their heads topped with red-and-white-checked headdresses or small prayer caps. Loose dresses swung from the women, their faces joyful, some inked with tattoos around their lips, like strips of lace.

A middle-aged woman in a black scarf and green dress touched my arm, as if to steer me somewhere. "You must go into another room and change your clothes," Levon instructed. My wrinkled brown linen skirt and cream shirt, my backpack's finest, apparently did not suit the occasion. The lady introduced herself as "Hala." She was the granddaughter of Sheikh Hammud al-Aekleh, and, I have changed her name, as I have for most of the members of the family.

Inside the house, Hala opened up the closet and selected a beaded burgundy dress, her pretty twenty-something daughters nearby, black liner accentuating their brown eyes. I raised my arms high and she slid the gown over me and pulled my hair back gently. Next, she wrapped a yellow scarf around my head. *"Jameeleh,"* she said, meaning "beautiful." I glanced in the mirror, hardly recognizing myself in this local Arab dress — I'd been made over, just like my grandfather.

Hala ushered me into an adjacent room, their formal sitting area where people waited. Some stood along the walls, while others leaned on armrests, or sat on the floor with their legs crossed and pea-green pillows at their backs. They all parted to allow me to pass. In my fancy gown, I felt like a debutante being presented to society; I waved and smiled, and the greeting came back a hundredfold.

I sat beside Antranig and Levon, not far from the Anizah Bedouin in his black-and-white clothes, distinct from the roomful of Sunni Arabs. How strange, I thought, that these two groups had been at war during my grandfather's time here. Everyone stared at me as a few chatted quietly in Arabic and someone sneezed. I didn't know where to start. But I had to; the room felt heavy with anticipation. At last, turning to my translator, I said, "Levon, please tell them, 'My grandfather had nowhere to go, and your family took him in. I never thought I'd find you. I have come here to thank you.'"

"We remember him," said the sheikh's grandson Youssef, sitting across from me. He was the elder of this clan. Though titles like his were mostly honorary now, they still carried weight. "Your grandfather was lovely. He was full of joy. He was honorable. He did his job with honesty. He was living like the people who lived here."

I scribbled this down in my notebook, trying to keep up with the dialogue as another man interjected, "We used to call him Gharees. He was handsome."

Gharees means "implanted" or "embedded" in Arabic, a fitting name for a transplanted foreigner. But *handsome?* I wondered. Especially after everything he had been through? He was many things — distinguished, intelligent, funny. These other adjectives seemed to suit him better. Was he remembering another deportee? It was possible. Youssef smiled and shared some more. "They wanted your grandfather to marry one of Ali's daughters, but your grandfather said no."

"He loved it here," I explained apologetically, "but he wanted to get home to his family."

Flashes of light burst, one after another. Teenage boys stood near the entrance holding their cell phones high, snapping photographs with their built-in cameras. Only now did I realize I was the only female in the room. In the doorway, Hala was almost lost among the many women, layers deep. I beckoned her inside. She hesitated, as if it weren't the custom, but I was insistent. Perhaps the men also gave their approval, because her round face brightened and her eyes sparkled as she squeezed through the crowd. We hugged and held hands and listened to Youssef continue his story, "The Armenians would walk near here. In our village, there were four Armenians at that time."

"Yes, it's exactly what my grandfather wrote," I said, before correcting myself; including him, it was five.

"Why did the sheikh save my grandfather and the other Armenians?"

"It is the teachings of Islam to be generous," one great-grandson explained.

"For the rest of his life, my mom said, he spoke about Sheikh al-Aekleh; he could never forget his kindness. My mom wouldn't be here if it weren't for that kindness. I wouldn't be here."

Curious, they wanted to know more about my grandfather's life post-sheikh. His number of children, grandchildren, great grandchildren . . . I traced a small family tree onto paper before proudly declaring, "We are seventeen!"

Levon translated this into Arabic, and I waited for awe to sweep the room. Not just one person saved but *seventeen.* Instead, their faces furrowed. "What a small family you have," Youssef said, as if he were sorry for my grandfather and his modest number of descendants. The poor guy — Stepan had defied the odds to continue his bloodline and all he had was seventeen! By comparison, Sheikh al-Aekleh had three hundred to five hundred descendants. No doubt half of them were packed into this room with me now, and spilling out the door, the robust lineage due in part to the tradition of multiple wives.

A young man entered the room holding a small pot of tea with a long handle and a single cup. At one end of the room, he began to serve the guests, working inward, constantly refilling the glass. At my turn, I swung it back, like a shot, possibly a little too practiced from my college days. The taste was warm and sugary, and I gave a big thumbs-up just as my scarf slipped off my head and my hair tumbled forward. I promptly fixed it as they all laughed.

Returning to my grandfather, I asked what else they remembered. One blurted out that he had blue eyes. I corrected him and began to worry again that they weren't talking about my grandfather Stepan. Just then, someone added, "He swam like a fish." Yes. I breathed easier; that was true, well known in our family. As the conversation evolved, I learned Sheikh al-Aekleh's son Ali had been born around 1910, and I felt sure again; that was within three years of my grandfather's recollection of the boy's age. I trusted I had the right clan, given the exhaustive search at Sheikh al-Ghubein's house, but knew that memory could be tricky and that the experiences of their rescued Armenians might have merged after so much time.

To fill in any gaps, I began to tell them more about my grandfather's life there. At first, I paraphrased his experience, but then I stopped myself. I wanted him to narrate. So I opened my lightweight Sony laptop and pulled up a translated version of his journals. When I recounted how his clothes were stolen by the flirtatious girl, someone said, "I heard that before." With another tale, it was the same. After all, this was their history too, I thought. Like folklore, his stories seemed to have been passed down through the generations, sometimes intermingled with others', sometimes not. Storytelling was very much alive here, I learned. As I read aloud, the packed room kept erupting with laughter. Even a century later, my funny grandfather was still able to entertain a crowd.

When I told them of the kissing at the river's edge, one grandson joked, "We will find you a husband too! Then you will never leave." My dating issues suddenly seemed solved.

Throughout it all, I had this unearthly feeling of knowing the importance of the moment as it unfolded. Since I'd begun this journey, something had happened to me, I realized. My life, my grandfather's, and my mother's had all become intertwined, like a wreath, and the circle was now closing. Though barely sunrise in Los Angeles, I called my mother, wanting to share all this good news. As soon as she groggily answered, I said, nearly breathless, "Mom, remember the sheikh who saved Baba's life?" "Yes, of course, Dawn." "Well, I found his family! I am with his family now." "What?" she squealed, then summoned my dad. "Jimmy, come quick!" On the crackly speakerphone now, she thanked everyone for their kindness toward Baba. As I listened, I found myself crying, for all the people lost, for all the people saved. Crying for the beauty of this family and crying for the determination that allowed me to step through time and find them again.

After we finished, Hala grabbed my hand and led me along an outside corridor, back into the heat where the women and children congregated. Giggling teenage girls climbed up an unfinished staircase on the exterior and sat in a descending line, bright faces atop staggered stairs, and gazed down on me below. Every so often, a woman would grab my shoulders and kiss me hard on both cheeks, her facial tattoos, a hallmark of beauty here, lifting with her smiles.

Inside the kitchen, I was shown a huge cauldron, the flames underneath curling up the sides. A young woman stirred the gelatinous liquid using her whole body, the ingredients unclear. This was our meal, she explained. With delight, she lifted up the ladle. Startled, I almost took a step back as the head of a goat emerged, its mouth open, its ears flopped over, its eyes like two cloudy marbles.

The animal had been slaughtered in my honor. Learning this, I felt grateful for their generosity as well as the animal's sacrifice. However, my squeamishness around meat had been lifelong, rooted in my California childhood of steamed vegetables and tofu. But I couldn't refuse this kindness. Nearby, a man began to spoon white rice onto a large circular silver platter; another man leveled the surface before carefully pouring on the stew.

After changing into a more casual traditional dress, I was led to a patio. A smiling boy brought out a bar of soap and an orange plastic watering can, and he poured the water over my hands to wash them. Someone draped a plain blue sheet over a multicolored Oriental rug for our dinner table.

In my grandfather's time, the most important people sat down first, and today I had that honor. Pointing to the pewter-colored platter, someone said, "This was Hammud al-Aekleh's. This is the same plate your grandfather ate from." *Incredible.* The intervening years washed away as ten of us sat around the dish. Atop the rice and meat was the goat's head, its teeth bared, its expression almost bemused, and beside it were stacks of breads and a tomato and cucumber salad, a veritable feast. With Hala next to me, I reached in with my hand, like the others, and tried to bring the meat and rice to my mouth. Everyone seemed to be watching me, including the goat, as half of the stew slipped from my fingers; the grease streaked my arms, and rice dotted my blue dress like miniature footprints. Not deterred, I tried once more, and dropped another handful of food. As my face heated from embarrassment, I remembered my grandfather's first attempt at eating with his hands and smiled at what was clearly a genetic predisposition for clumsiness.

Home

1918-1919

DR. GARABED rushed the injured soldier down the Jerablus platform, nearly carrying him while shouting, "Help him get on the train!" A sergeant attempted to lift the feigning Stepan but failed. He motioned for assistance. Two more subordinates appeared and hoisted him into the car. Stepan's eyes adjusted to the scene inside: ailing fighters everywhere, sallow and emaciated casualties of war. Their moans scaled into a chorus, and he added his cry.

"This one won't last much longer," some whispered around him.

In ten minutes, the whistle blew down the tracks. All night, Stepan encored his performance as both he and the train screamed southward, past indecipherable stations. In the morning, they arrived in the noisy streets of Aleppo, with its renowned souks, minarets, and medieval citadel sweeping the sky. Commanded off the train, Stepan joined the crowd that lined up to eat and reached for his ration of bread. Weak soldiers needed their strength, as did the impersonator in their midst.

"Get back into your cars," bellowed Dr. Garabed an hour later. The men snapped into line for the head count. "Two . . . four . . . six . . ." With an air of routine, the physician added Stepan to his tally and boosted him into the car. Another sergeant picked up where he had left off, "Ten . . . twelve . . . fourteen . . ."

The train chugged north and then west into a fading sun. The scenery shifted. After Qatma station, they passed mountains and a wooden bridge called Tahta Köprü. Stepan had crossed it by foot with his labor battalion three years earlier. In fact, he was retracing his journey. Since he'd first come this way, so many had perished before his eyes.

After switching train lines, Stepan and the sickly cut across a valley, past the settlement of Ishlahiye. Boring through the Infidel Mountains, the locomotive raced through the recently completed Baghche tunnel. When Stepan had trekked through here in 1915, he and his battalion spent days scaling its peaks, passing looted Armenian shops, the dead strewn at his feet.

The train emerged from the darkness, and the soft light of early morning filtered into Stepan's compartment. Looking out, he recognized Mamure, on the mountain's other shoulder, studded with trees. It was a marked change from the baked desert of Mesopotamia. Stepan had also toiled in this area, hauling logs down hills with his comrades. Much of the tunnel's hard labor had come from the Armenian pack animals, like him, and British and Indian prisoners of war who had helped burrow through the pass that was blocked with quartz.

Told to disembark, Stepan waited for his rations in the company of new men, conscripts from other cars. Their hungry eyes narrowed and fell on him: The stained and ripped fatigues. The shoeless feet. The uncovered head.

"You're a stranger," they said.

"I'm not a stranger," he protested. "My papers are with Dr. Garabed." Then he learned some crippling news: the doctor had accidentally been left behind in the last transfer. That meant Stepan's forged documents had been too. His story, the thread of his disguise, was unraveling. He had to do something or he'd be apprehended. Around him, the mob pulsed with suspicion, and some even threatened to alert the station police. The distribution of rations distracted them, as ravenous men thrust out their hands to receive the slaps of bread. Seizing the moment, Stepan squeezed through the bodies and disappeared into the bushes. Crouching low, he waited until the train started down the tracks. He was stranded, without food, papers, or money. At the foot of the rise, he spotted a restaurant and soon groveled inside. "I am ready to do anything, as long as you would let me eat whatever leftovers there might be on the plates . . ."

"What are you doing here, Stepan?" asked a man from Adabazar named Hovhannes. Stepan was dumbfounded. Like Stepan, Hovhannes had wound up in this rocky purgatory after the deportation, assisting the cook in order to survive.

The next few hours unfolded seamlessly. With a recommendation from Hovhannes, the boss hired Stepan to sweep and scrub the floors, fetch water, knuckle down to any task, all for squiggles of scraps. He spent perhaps the next few months this way. He was biding his time until his next step, he told himself, which would certainly take him toward home.

One morning, a customer entered, hungering for *madzun*, yogurt. As it was out of stock, Stepan hurried to the market to replenish the supply. His brisk walking and fatigues, however, attracted the attention of some police officers. "What are you doing here?" one hollered. "You're a soldier."

"No, I'm not a soldier. I'm a refugee." He gripped the tray in his hands.

"Then what's this you're wearing?"

"I had nothing to wear and a soldier friend of mine gave it to me because they were worn and old," he said.

"Ahmet, arrest this soldier."

Stepan submitted, and they manacled his hands and marched him to jail.

All over the theaters of Europe and the east, fighters were deserting in droves as the war ground to a close. In the Ottoman military, more soldiers had fled than remained, leading the Germans that June to contemplate withdrawing their troops from their joint campaign in Palestine; an unsuccessful German offensive against Allied forces in France and Belgium had already weakened the Central Powers. Still, the Ottoman leaders struggled to hold on, with their exhausted troops and dwindling supplies, to the land that had been theirs for more than four hundred and fifty years. In the conflict's twilight, the Ottoman government directed scarce manpower toward taking Russia's old Transcaucasia provinces — Georgia, Azerbaijan, and Armenia — which had become independent after the revolution. Creating a special "Army of Islam," the war minister sought to reunite the Turkish-speaking people of Asia Minor with those in Azerbaijan. The lands of Armenian settlement from Eastern Anatolia to the Caucasus sat between the two, between this pan-Turkish and Muslim dream.

In jail, nine men, similarly accused, shifted around beside Stepan. Gendarmes unlocked the door and escorted Stepan and his group to the station. "Where is the train heading?" he asked.

"To Adana."

Good, he thought, *toward home.* Adana wasn't that far west, but it was in the right direction. Stepan knew those roads, was confident he could find one on which to slip away. His ability to flee had become an ingrained instinct, second nature. To flee, always to flee, dominated his thoughts. But it wasn't time yet. He rested easily as his train clattered across the Cilician plain, speeding away from one range, the Amanus, toward the slopes of another, the Taurus. The other passengers had also been humbled by war. The sacrifice of the last four years had been high for the Turkish army; many soldiers had been lost on the campaigns of Mesopotamia, the Caucasus, Sinai, Palestine, and Gallipoli. Of the enlisted, the fatalities would reach over seven hundred thousand.

The morning revealed the town of Osmaniye, a onetime Armenian camp. In another thirty or so miles, the men climbed off the train, in Jeyhan, near a military base. Stepan wasn't going to be killed for desertion. With the ongoing conflict, the military badly needed men, regardless of creed, even Armenians. Onward toward Tarsus, the group marched, and a lovely flour mill emerged, perched on the banks of a tributary. It was a sprawling operation, a series of one-story buildings with a gazebo, where the water flowed and skated over stones and into small pools of white foam. This was their destination, he learned, and they were here to work.

Inside the mill, sacks of grain rose into towers that grew by the moment as laborers carted in more staples. Wheat. Corn. Barley. When had Stepan last seen this much food? Assigned to be a porter, he lifted and weighed the goods that were then ground down into a fine powder. The morning disappeared. A soldier arrived and read aloud names from a list, then placed his hand into a sack and magically withdrew a whole loaf of bread.

Lunch had almost surprised Stepan. He had forgotten the regular world and its routine of meals. He wouldn't run away, at least not yet. This place was different; the mill's owners were Armenians, exempted from the deportations in order to supply the military. At one time, Chalvarjian brothers produced around 121,000 pounds of flour each day for the estimated 25,000 soldiers stationed in the area. Secretly, Mardiros and Khacher Chalvarjian had also used their factory for another purpose—to shelter countless Armenians, from some of the most famous intellectuals to, now, Stepan.

One afternoon, after placing fourteen wheat sacks on the scales, Stepan loitered while the clerk tallied the figures. Curious, and perhaps a little bored, he peered over his shoulder and scanned the math.

"It's wrong," Stepan said. "It's two hundred kilos short."

The clerk repeated his calculation and then ushered the porter to the manager, Haygazun Efendi. After studying him a long while, he finally asked, "What school have you graduated from, my son?"

"I went to school until I was ten years old. When my father died, I left school to work."

The *efendi's* eyes darted between Stepan and his clerk, as if the look on their faces might solve the mystery of this uneducated man's mathematical abilities. Handing Stepan a pencil and paper, the *efendi* recited fifteen numbers he wanted totaled. Stepan completed it in less than a minute.

"Are you sure you added them correctly?"

"Yes, because I'm very good at calculation." He'd had to be, with his years on the streets as a peddler and courier.

Haygazun consulted the figures—they matched with his total. At once, he took Stepan to military headquarters. "I needed an assistant clerk, and now I have found one," he told his superior. Stepan's promotion was effective immediately.

In addition to a fresh set of clothes, Stepan received a new hat and his first pair of unworn shoes in four years. No longer itchy, he felt clean, his feet protected as he walked. Alongside a team of Armenians, he worked in the firewood division, enjoying regular meals as well as a paycheck: ten liras in paper money at the top of each month.

He penned a letter to his family, using their last known address: Bolvadin, the village they'd gone to after Chai. Despite his fortunate situation, he still felt compelled to follow the letter west, but he was no longer sure of his motivation. Was it his longing for them, he wondered, or had his urge to flee become a habit?

After receiving a response from his family, he wrote again, the back-and-forth breathing new life into him as the war's end seemed to near. In Syria, with Arab troops obstructing the railways, the Allies had pushed through Ottoman defensive lines in the Battle of Megiddo, and were marching toward Damascus. Bulgaria succumbed to a French offensive that September and signed an armistice. Out of the war, the Allies pushed toward a new front, on the Danube River, leaving the Germans and the Austrians vulnerable from the south. With the mounting defeats, the Ottomans broached the topic of peace.

At the flour mill, Stepan studied the shipments fanning out across the soon-to-be-defeated empire. Every few days, officials accompanied the barley flour and corn to their endpoints. His chance for escape, he realized. Though initially rebuffed, he pestered his superiors until they finally relented and allowed him to accompany the grain. They dispatched him to Bozanti, a town on the opposite side of the Taurus Mountains that would put him that much closer to home.

In his boxcar, he lay beside the pillows of barley flour. As the train moved out of the station, he promised himself that he wouldn't return. He would go in only one direction now: toward home. However, the train languished in places for days at a time. First in Yenije, then at the tiny mountainous station of Dorak, where he was now stranded. From his soft, dusty bed of barley flour, Stepan glimpsed the world through the open door as the nutty aroma of grain wafted outside and through the terminal. Peasants poked their heads in and asked to buy the grain, but Stepan refused, despite the temptation of cash. After Stepan overheard Dorak's stationmaster speaking Armenian, he befriended him, and the pair soon whiled away the hours reflecting on the sufferings of their people.

The next morning, as Stepan dozed, the heavy boxcar door rattled open, startling him awake.

"Who's there?" he asked in Turkish, thinking it was another pesky peasant.

"Hurry up and come to my office," the voice said. It was the stationmaster. Sleepy-eyed, Stepan scampered up and followed the stationmaster down the platform.

"*Achki luys,*" the stationmaster said in Armenian. It meant "May the light of joy shine in your eyes." He gripped Stepan's hands. "Congratulations — an armistice has been signed! I just received a telegram."

"Glory to God! We've survived!" Shocked and ecstatic, the two had dreamed of hearing this news for four long years.

German convoys soon appeared, one after another, as well as lines of Turks, fresh from battlefields. Like a river that suddenly changes course, the animals, the soldiers, the cars, everything that had been needed for a war were flooding back toward home, choking the roads.

As the Turks retreated, soldiers destroyed their ammunition and supplies, creating great blazes that marked their trail. Damascus had fallen. The British and Arabs had seized the city on October 1, 1918. After this, Beirut followed, then Aleppo. The spiral prompted the cabinet of the Young Turks to resign. The men who had come to power on a platform of equality and then orchestrated one of the worst ethnic slaughters in modern history had stepped down.

On October 30, 1918, the Turkish war seemed to be over. On a boat near the island of Lemnos, the government surrendered and declared a cease-fire with the British. Hostilities ended the very next day. Under the terms of their armistice, the Turks had to disarm, and the Allies could occupy any area if needed. Nearly two weeks later, after the collapse of the Austro-Hungarian Empire, the Germans signed an armistice with the Allies, officially concluding the Great War on November

11, 1918. All told, approximately ten million had died and twenty-one million had been wounded in this war that was supposed to end all wars.

With news of the war's conclusion, Stepan was nearly bouncing off the flour sacks, his excitement ricocheting around the boxcar. On the other side of the Taurus Mountains, his train stopped in Bozanti, the destination of his grain. *I'll be free and get my papers,* he told himself, his nightmare almost over.

But his service was far from complete. Ordered back to base camp and saddled with another flour delivery, he spent the next three agonizing weeks shuffling the short distance between Adana and Tarsus as Allied forces swept deeper into Ottoman territory. Everyone seemed to be bound for home, everyone except him. Maybe his family was too. He didn't know if they had left Bolvadin.

The sea around Constantinople became walled with Allied warships as leaders began to debate why the country had gone to war and who was to blame for the Armenian atrocities. Calling for trials, the lawmakers nullified the temporary decrees that had legalized the Armenian deportation and the sale of Armenian property. Soon after, the parliament was dissolved yet again, and a new one formed.

Afraid of prosecution, the former leaders of the CUP vanished. Talaat Pasha, the grand vizier; Enver Pasha, the minister of war; and Jemal Pasha, the minister of the navy — all of them disappeared beneath the waves in a German submarine. "Where have the *pashas* gone, running from place to place with cudgels in their hands, scimitars in their belts, blood in their eyes?" asked an Istanbul newspaper.

Finally, Stepan pulled in with a load of flour and asked for his discharge papers. Because he'd turned in the shipment intact, because he'd been honest, because he stood out from his predecessors, he received a month's rations and twenty liras. Astounded, he stared at the certificate and the money in his hands, the most since his courier days. Although burdened with another eight cars of flour — truly, his last — he possessed the money *and* the papers to make it home. Now he could see his family, if he could find them. At the bazaar, he purchased ten sacks of oranges, separated the fruit into bundles for resale, and boarded the train westward, toward Adabazar, gambling that his family beat him there. The whistle shrieked. He felt like singing.

It was already December; 1918 was winding down. Back in Yenije, Stepan's train temporarily halted. Overhead, the skies welled up, and poured down on Cilicia, as if to cry for all that had happened. The mountains, the trees, the rooftops, all soaked with rain.

At the time, the newspapers were filled with news about their fugitive leaders. An extradition order had been issued for Talaat Pasha, but Germany refused to turn him over. Nevertheless, punishment for the Armenian crimes seemed imminent. "Sultan Searching Out Authors of Killings" reported the *New York Times,* and "Millions Slain by Turks." As the Ottoman parliament was officially dissolved, another headline promised a "Court-Martial to Try Officials Responsible for Massacres."

Moving out again, all Stepan could think about was getting warm. No matter what he did, he was bitterly cold, chilled, and feeble. Sitting felt torturous and he could barely keep food down. He, who had crossed deserts and mountains by foot, grew so ill he couldn't stand anymore. *Am I going to die before reaching home? Will I be left on the road?* All over the world, many others were wondering the same thing. To survive the war, only to confront a new enemy: the Spanish flu. Although Stepan didn't know what ailed him, the influenza virus was quickly becoming the worst pandemic in history, spreading among close-quartered servicemen in the Turkish army and elsewhere. Eventually, it would overshadow even the bloody conflict, with twenty million dead by its end.

As the train stopped and started, Stepan drifted in and out of consciousness, his progress home so achingly slow, he began to weep. He arrived at Dorak station, where he'd first learned the war had ended, and realized he'd barely traveled any distance from his starting point. Near the tracks, masses of people were gathered, jubilant with the news that the French had captured nearby Adana, lowered the Ottoman flag, and raised the French one. It was December 21, 1918. Stepan should have been ecstatic, but instead he spent the night in a feverish haze, feeling sorry for himself. The next thing he remembered was a whistle blasting. Morning already, and another train was squealing into the station. Now, a bang at the door: *Tuk, tuk, tuk.*

"Who is it? What do you want?" Stepan asked.

"The French have come and occupied the station," said the voice. "I want to have some flour."

He refused, but this news revitalized him. Easing himself down onto the platform, still not far from his goods, he could see servicemen spilling out from the other locomotive. He scooted closer, unable to believe the sound.

"Are you Armenian?" Stepan asked.

"Yes, we are. All of us, all these soldiers are Armenian. We're taking over the entire line, all the way to Bozanti. Where are you from?"

"Adabazar," Stepan said.

Immediately, they called out, "Vahram, come, there's a compatriot here." Men came running. Stepan was astonished at the coincidence: two were friends from home.

"What are you doing here?" they asked.

"I'm going to Adabazar. I've got discharge papers."

"Oh! Wherever we go is considered captured. We're going to Bozanti to capture it as well. Take everything you have and come with us. We'll take you to Bozanti."

As the Turks withdrew west, the French pushed into Cilicia. Stepan's long-lost friends were tasked with liberating the area and staking the claim for the French. Once part of the Légion d'Orient, the Armenian volunteers had already assisted the British in Palestine and were steadily moving inland. Many were survivors of the massacres themselves. Suddenly in a position of power, some went to take revenge on the Turks, an unfortunate sequel to their suffering.

But in the moment, the idea of joining the Légion was irresistible to Stepan. *I, who had been the conquered, will now become the conqueror!* As he boarded the train, the French captain sized him up quizzically.

"*Qui est là?*"

"*C'est un Arménien,*" his friends replied. He is an Armenian.

"*Très bien, très bien,*" he said. Finally, someone was happy to see him. Jubilant, the soldiers rode farther into the Taurus Mountains, where

Stepan had first realized his life was in danger. Reaching the other side, the soldiers were greeted by loud music, as the railroad workers of Bozanti serenaded them with Armenian songs. Stepan listened, his spirit bounding despite the fever and the chills racking his body. He would recognize the music anywhere, these notes of a people nearly extinguished.

Finally, at the limit of the Légion arménienne's jurisdiction, Stepan disembarked. He straddled Anatolia now, Turkish territory. Some two hundred yards away stood another train of Turkish soldiers, headed farther west. Discharge papers in hand, he approached the officials. The nearby cars were mobbed; men covered every bit of floor and rooftop space, even in this frigid air. Undeterred, he presented his papers.

"There isn't room on this train. You take another train tomorrow."

"It's cold, how can I stay here?" he asked.

The man didn't answer. It was just before New Year's Day 1919; and Stepan's health was fragile. He was also stuck at the foot of the snow-capped range. Hundreds of other servicemen were similarly marooned. He sneaked forward anyway, searching the train for any spot to sit, his goods a priority and already onboard. In the first boxcar, nothing. In the next one, the same. There, in another, lay a very sick fighter, cold as a stone and uncovered.

"My friend," Stepan said. "I have a *tente*."

The man did not respond, even to this offer of a cover to warm himself. Cramming in anyway, Stepan unfolded half of the tarp across the man. Then he bought some *yershig* (sausage), cheese, and bread from a vendor. Thirty minutes later, the train edged out, with Stepan and the stranger pressed next to each other, Turk and Armenian sharing the warmth and the food. Onward to Ereghli, where he delivered his goods. At the next stop of Konya, the gendarmes scoured the cars for the sick and the dead. Around Stepan, seven men had already died. "Disembark for inspection," someone ordered.

Stepan worried he'd be singled out and thrown off. Summoning strength from some unknown reservoir, Stepan rose to his feet and hustled toward the back exit. He was almost there when a guard intercepted him.

"My friend," Stepan said, standing straight as he could, "the inspection is for the sick; I'm not sick." He passed him a rolled cigarette, along with one lira. "Allow me to take care of my 'natural need.' I'll be back shortly." His usual ruse. He ambled away and squatted down. Then he strode back to the car, erect as his compact frame permitted, and waited until the purge of fifty men was complete. The train screamed out without them. Heading west, Stepan was free, finally free.

During those cold January days, the peace conference opened to great fanfare and expectations. Gathered together in Paris, the delegations of some two dozen nations began the daunting task of finalizing the terms of peace as well as finding a way to prevent war from ever happening again. To protect smaller countries from belligerent ones, they spoke of creating a new international body called the League of Nations.

Victorious France, Britain, Italy, and the United States presided over the negotiations as all sorts of wrangling broke out over reparations and the drawing of new borders. The division of the vanquished Ottoman Empire caused particular friction. What should happen to Turkish Armenia? Should there be an Armenian mandate so the people could be safe from future persecution, with the United States as overseer? And how about Palestine? The British sought that mandate. Syria? The French and the British both had interests there. And Italy wanted its share. Everyone referenced all the diplomatic commitments signed during the conflict, and the European powers believed that they could govern the people better than the residents themselves, despite the Arabian king's proposal for an Arab federation that would oversee Yemen, Mesopotamia, Nejd, Syria, and al-Hejaz.

To complicate matters, none of the Central Powers were invited to the deliberations. Neither was Russia, a onetime ally, given its ruling Bolshevik government. The process dragged on so long that some worried the weakened Turks would regroup, setting off another battle for Anatolia.

On the train, Stepan and the soldiers traveled westward across a windswept landscape, the days collapsing, until they eventually veered north.

In the evening light, he finally pulled into Arifiye station, just five miles from Adabazar. Stepan joined the exodus onto the platform and made his way to the front of the building. With connecting service suspended, a line of horse-drawn carriages waited, the drivers shouting in the twilight, all begging for a fare. Wan from illness, Stepan examined their faces, their dress. All Turks, he realized, and he proceeded with his search. "Stepan Akhbar, I have a carriage," offered a boy in Armenian. "I transport people to Adabazar." Stepan recognized him, the son of Lusi, a water carrier. At least, that was what she had done before the deportation. Stepan climbed in beside another serviceman and began the final leg home.

The wooden wheels turned as the winter night descended. With each mile, he felt that much more alive. His illness seemed to be lifting, even though he was unsure what he'd find. He didn't know if his family had survived the final days. He didn't know if they'd made it home. An estimated 1.2 million Armenians were believed to have died. Later estimates were higher still.

The boy held the reins steady, steering the carriage into the municipal limits of Asdvadzareal Kaghak, the God-Created City. Or was it? Stepan swept his eyes around; the images, painfully, did not match his memories. The horse trotted past the bazaar and turned onto Kara-aghach Dibi, Under the Black Tree's main street, once the center of Armenian life, the route of the tragic alphabet procession, the home of the wonderful bar Manol's Sea, the Massis Theater, the Getronagan School. "Where has my cheerful city gone?" he asked himself. At this time of year, the weather was chilly. Even in the dark, he could tell no one was pacing the streets. Piles of ashes stood in place of buildings. All that was important to their community — their churches, their stores, their homes — desecrated. One house of worship had been turned into a stable, another into a warehouse, its belfry ripped out. Even the cemetery had been denuded of its gravestones. A part of him felt erased. Not far away, he could see a flicker of life: a few merchants were selling goods by candlelight.

Renaissance, an Armenian newspaper, reported in December 1918 that Adabazar was "a mass of ruins" and that a "large part of [its] houses had been demolished." A telegram sent to the region's leaders permitted the Armenians to return to their villages, yet many soon discovered Turks squatting in their homes. Some had even taken two houses, one to live in and one to strip for firewood. A few Turks refused to evacuate, and the livid Armenians beat them to death. The body of one man was left outside, like a scarecrow, as a lesson to the rest.

Stepan's carriage halted. The other passenger paid the driver and hopped off. "Drive on!" Stepan ordered. "Drive on to our house!"

The horse tugged them onto a wide avenue and then turned onto another path that led past the girls' school. All rooms were dark, the pupils and books and American missionaries long gone. He rode into his neighborhood of Nemcheler and onto his street, not talking, just thinking, looking, smelling the air. The driver stopped in front of the Miskjian home, where silkworms had once curled around mulberry branches. At least it still stood.

Stepan handed the boy a lira, eagerly tipping him the twenty ghurush in change. He stepped out and placed his feet down firmly on his street for the first time in four years. As the carriage clopped away, Stepan studied the buildings, expecting to see signs of his people. But there was nothing, only a panorama of blackness. On the first floor of his home, a dull light glowed. One of his sisters? Or a trespasser? He walked to the door and rapped his knuckles on it — *tuk, tuk, tuk* — his heart beating the same sound. No movement from inside. Standing there, he heard the huffs of his breath. He knocked again, harder this time. He couldn't wait any longer. The window on the top floor creaked open.

"Who is it?" yelled his sister Arshaluys.

"Stepan, your brother."

All at once, he heard screaming and his name — *Stepan, Stepan* — and the scramble of heavy footfalls down the stairs. Above the commotion, the sweet pitch of his mother's voice rose to the top, "First me, first me, first me."

The door swung open and there stood Stepan, the missing brother, the missing son. He was emaciated and wearing only a few shreds of clothing, deathly ill. He, too, examined them and counted each one. *My family is alive*, he rejoiced. His dainty mother swept him into her arms, the tears streaming down her cheeks. His siblings followed: his big brother, Armenag, and all four of his sisters, Arshaluys, Aghavni, Mari, and Zaruhi, Ovsanna's mother. One by one, they reached and hugged and held him tight, marveling that they'd all survived.

His family had been through much strife too. When they'd first approached the main family home after an absence of more than three years, black markings striped the window — not the way they'd left it. Discovering a family living there, they demanded the intruders leave, taking control after having none for a long time. The people, mercifully, heeded their order. Already, Zaruhi and her daughter, Ovsanna, had seen what had become of their own three-story house. Most of the buildings in their Armenian neighborhood had been razed. From afar, they could see theirs had been too.

Now, in the doorway, Stepan's mother raised her hands to the heavens. "My *murad* [wish] has been granted. Glory to You, my Lord. I got my wish. Now I am free, You can call me. Call me whenever You wish so I can come near You."

She could die now, she didn't care. Her family was intact. Ovsanna remembers this day, when the family, for a brief moment, became whole again and tried to move forward.

Stepan struggled with the enormity of the losses, floating through the house like a ghost. "For a while, he wasn't able to speak," she said. The faces of the dead seemed to be parading through his head, the screams of the dying. Somehow, he'd escaped the desert, but he would never escape the past.

One Family

2007

FROM THE PLATTER, the dead goat continued to watch me eat. I was spilling food everywhere; still the sheikh's descendants piled more onto my plate. "I'm so full," I said, yet they shook their heads and repeatedly protested in Arabic, *"La, la, la, la."* No, no, no, no. Like my grandfather, I hadn't consumed this much since leaving home and was stuffed after a few bites. I tried to appease them with one last handful, but the meat again slipped through my fingers. Finally, my hosts laughed and brought out a spoon. Defeated, I accepted it, and they shared more about Sheikh Hammud al-Aekleh. I learned he had died in the 1930s, at around age seventy, and was buried nearby.

"I'd love to pay my respects. Can we go?" I asked.

When everyone had finished, one of the sheikh's grandsons hurried me into a car, and we trailed a line of vehicles toward the bluff and the cemetery at its base. In the early-evening light, the shadows fell stark and long, the dusty ground covered with stones. Unlike the American graveyards, with their manicured lawns and marble headstones, in this cemetery, simple stacked rock configurations and tombstones marked the dead. Most of the group walked ten paces ahead, the men's *thobs* and long headscarfs blowing in unison behind them. Cautiously, they stepped past the stone markers, some askew or fallen flat as if bent in prayer. I followed in silence, the crunch of footsteps echoing behind me.

A bearded man ushered me to a small headstone in the center of the cemetery. It belonged to Ali, his grandfather, he explained, the sheikh's brother whom my grandfather had fed the curative *madzun*. I bowed in obeisance, bending my knees and lowering my head, as another man's voice called out.

"That is Sheikh al-Aekleh's," he said proudly, and he pointed to one of the tallest tombstones, scoured by the desert winds, awash in fading green paint. I stood there for the longest time, thinking about this special man who had understood my grandfather's true identity and opened his heart and home. After all this time, I'd finally found him, and I felt a lightness in my chest as the burden lifted, the realization overwhelming me. It was more than what I could have imagined. I leaned in and gently kissed the stone, cool against my lips. "Thank you," I said, just under my breath. "Thank you for saving my grandfather's life."

The sun was beginning to set, but I hadn't visited the Euphrates yet. A dozen of us piled into the bed of a pickup truck and headed toward the river, following a caravan of cars. Slicing through green fields, with the warm breeze blowing against my cheek, I remembered how someone back home had asked me if I had any preconceived notions about this part of Syria, with its strict control by President Assad, and if I was worried for my safety. Looking at the smiling faces around me, of the old and the young, I felt ashamed that I had ever felt any fear.

For years, I'd wondered how my grandfather had survived. After traveling thousands of miles, I'd finally uncovered the reason, and it could be found beside this riverbank. The best of humanity existed here. I thought about the Bedouin sheikh of Raqqa who had hosted me like a daughter and helped to locate this family; the strangers who stopped what they were doing when Levon, Antranig, and I were lost and set us in the right direction; my translators and drivers; and the altruistic family friend in Aleppo who had accepted me into her home, sick, and nursed me back to health. I thought of the turn of events I'd been through to get to this place — moving home to Los Angeles; finding my grandfather's journals, a discovery that felt like he was calling me; crossing the border alone. I also thought of the Turks I'd met along the way: the two women who shampooed my hair at the Kilis hamam and the nice Bolvadin taxi driver who had invited me to his relative's wedding despite my ethnicity, despite what he'd been taught about the Armenians.

At one point, Hala's son Omar told me that when his extended family had first learned I was coming, they'd thought I was in need, like my grandfather. It was their tradition, the twenty-seven-year-old explained, to help someone in crisis. They would feed me, shelter me, whatever I required to make me whole. They had welcomed both me and my grandfather in this same spirit, both of us strangers, with different ethnicities, languages, and cultural beliefs. "A friend of the clan will not be asked what religion he embraces," one of the villagers later explained. "Religion is for God; the homeland is for all."

As our truck rocked down the rutted dirt road toward the Euphrates, two shepherds cleared their slow-moving sheep out of our way, a job that had once been my grandfather's. At the marshy bank, I hopped down and gazed out at the serene river. "This is where your grandfather's clothes were stolen," someone said. I could almost picture him there, where he swam "like a fish" and wrestled with the coquettish girl for his garments until she relinquished them for a kiss. Some kids jumped into the water and started splashing around; some adults waded in too. Hala was up to her knees, holding her dress, standing beside her sister, her son, her nieces, and she beckoned me forward with a smile that filled her round face. I removed my flats and hoisted up my blue dress so it wouldn't get wet. The water chilled my legs, although the air was still warm. I waded out, feeling the mud between my toes. Cupping one hand, I dipped it into the river, and drank some of this refreshing water that had revived my grandfather after his desert crossing. The light was almost gone, the sky's shifting colors of orange and purple and pink seeming to dance, just like the water.

Small waves gently lapped against me, washing away the sadness I'd held tight in my chest like a ball of barbed wire after obsessing over the genocide for months and for years. In trying to reconcile my family's past, I'd lost sight of the future and allowed a mix of anger and melancholy to grow. Neither emotion heals, despite how justified they may feel. My grandfather had let the anger go, and so did I, letting it be carried away by the river. Sheikh Hammud al-Aekleh had thought for himself and had balked at prejudice. Just one person, yet he transformed the fate of my family. What would happen if more people in Turkey questioned their own history and faced it, no matter how painful or costly? What if they followed in the footsteps of Sheikh al-Aekleh? And what if more Armenians let go of their rage, which prevents some from even talking to Turks, and learned to forgive?

All around me were descendants of the family my grandfather never stopped thinking about his entire life, always regretting having left without saying goodbye. I felt so emotional that I could return and express my gratitude for him, for us. My grandfather also couldn't forget the brave nameless Turkish soldier who had battled his friend to spare an Armenian. He'd most likely traded his own life to save my grandfather's, and I wished I could have thanked his family, too, for his sacrifice.

Sopping wet, as the river flowed around me, I was truly happy. My thirty-six years had led to this moment, where the past and the present joined hands. I'd embarked on this quest for my beautiful mother; and two generations earlier, the same love of family had compelled my grandfather to go home. I yearned for my mother to be here with me now, dipping into the water along with these women and men, laughing and chattering in the twilight. I thought about her father and the persecution that he faced and the safe haven he found out here in the desert, with these people, within this oasis, flourishing at the water's edge. And I realized this was not just his adopted family but mine too, one big family united by an act of kindness nearly a century before.

EPILOGUE: WEDDINGS AND ANNIVERSARIES

"HAPPY GENOCIDE DAY!" I joked to my mother, as she stepped into my car in the spring of 2012. She laughed. As usual, we were using dark humor to get through a difficult day: April 24, the anniversary of the Armenian genocide. Already behind schedule, we sped east toward Montebello, an industrial city ringed by freeways outside Los Angeles, until we spotted the sign — *Armenian Genocide Martyrs Monument* — and a line of cars waiting to turn. This morning, hundreds of young people, the grandchildren and great-grandchildren of Armenian survivors, had marched down Hollywood Boulevard demanding that Turkey declare the massacres genocide. Now our more staid commemoration was about to begin.

"Oh, Dawn." My mother sighed. "We've left too late!"

"I don't think there will be a lot of people today," I said. "It's so cloudy. Plus, it's a workday."

As we drew closer, I could now see that the traffic was backed up far from the grassy hilltop, where the arched monument rose. The slender structure has been standing here since 1968, after donations poured in from the large diaspora for its construction.

In gridlock behind vehicles draped in Armenian flags, my mother glanced at me, her expression smug. "Oh, really," she said, as organizers directed us into another parking lot, "it's not going to be a big turnout?"

After parking, we hurried toward the monument, where we searched for two empty seats amid all the chattering people dressed in funereal black and dark T-shirts emblazoned with *Never Forget*. As usual, bumper stickers of *Boycott Turkey* were being passed around like mints. Scanning all the solemn faces, I asked my mother if my grandfather ever attended these public anniversaries.

"No," she answered quickly. "Every day he lived the genocide. He didn't need to come."

At the podium, a man appeared and belted out "The Star-Spangled Banner," with a thick Middle Eastern accent. A choir sang a mournful song, and a line of clergy in black robes led the prayers as the people began the annual painful process of remembering the dead.

In the wake of the massacres, many Armenians tried to forget and move forward. Of course, my grandfather couldn't do that. His thoughts and his conversations always drew him back to the desert, to that cistern of the dead. Still, that didn't stop him from trying to rebuild his life. Immediately after the war, the Adabazartsis wrestled back pillaged properties, reopened schools, and set up shelters for the refugees. Sickly after years of deprivation, Armenians received free medical care, daily bread, and weekly rations of meat from donations. Orphanages in Constantinople filled up with children from everywhere, including Adabazar. Three were Tevon's, the friend Stepan had buried on the road. Adabazar would forever have a gaping hole in its heart; only 10 percent of the Armenians ever returned.

Despite this, the Adabazartsis united, showing the world that their spirit hadn't been broken. At gatherings in the reconstructed community, a choir of eighty expressed themselves through song. Couples rushed to the altar and quickly started families as if to make up for all the loss. Sports also resumed, and at one track meet, my grandfather and his friend Harutiun Atanasian judged as the teens leaped and sprinted across finish lines. Not long before, they'd all been fighting for their lives. Happiness abounded with the announcement that Archbishop Stepannos Hovagimian had survived and made it back from the long road of deportation. The community marched to the railroad station to greet him, the band at the front playing instruments that they'd just reclaimed from the government. Triumphantly, they escorted the prelate to the Surp Hreshdagabed Church and, in the shadow of the missing belfry, showered him with flower petals.

Across ethnicities and faiths, the long process of reconciliation began. Not only in the Ottoman Empire, with its own wounded soldiers returning home, but around the globe. On June 28, 1919, Germany, along with the Allies and the Associated Powers, signed the Treaty of Versailles. Under the harsh terms, Germany had to accept blame for instigating the war, pay reparations, demilitarize, and cede much of its land to the victors. When the Paris Peace Conference finally concluded after another year, it did so without deciding the long-term fate of the Ottoman Empire. Syria, Palestine, and Mesopotamia would become mandates, governed by Britain and France, and the occupation of Asia Minor would continue, with the British overseeing the dispersed Greek, Italian, and French troops.

With pressure mounting from the Allies over the Armenians, the obsequious sultan had formed an Ottoman military tribunal and arrested some one hundred and seven people suspected of committing Armenian and other war crimes and atrocities. One of the first trials found that the local leaders had committed acts "against humanity and civilization." The provincial police commander of the *sanjak* of Yozgat was sentenced to fifteen years of hard labor, and the lieutenant governor to death by hanging.

At his funeral, attendees expressed their outrage and laid wreaths emblazoned with the words *To the innocent victim of the nation.* With the backlash following the lieutenant governor's death, the Ottoman government released some of the detainees who were slated for trial.

Though Talaat Pasha, the former grand vizier, and Enver Pasha, the former war minister, had fled to other countries, they were sentenced to death in absentia. "The disaster visiting the Armenians was not a local or isolated event. It was the result of a premeditated decision taken by a central body," stated the indictment. İbrahim Bey, the torturer at the Adabazar church, was also sentenced to fifteen years for similar acts in nearby Izmid, but he, too, had vanished.

Stepan waited for these men — and other perpetrators — to be held accountable, a wait that would last a lifetime and beyond.

The greater struggle for Anatolia was under way. My grandfather watched as forces loyal to the sultan moved into Adabazar to halt a rebellion that was sweeping westward from the country's interior. The Turkish people were expecting severe conditions postwar — the loss of Mesopotamia, Palestine, Syria, and control of the straits — but it was the landing of Greek troops in mid-1919 at the behest of the Allies that sparked the greatest dread; they feared that the Greeks would remain beyond their mandate and claim the land. An estimated two hundred thousand people protested in the capital, and General Mustafa Kemal, a Gallipoli hero, led a revolutionary party known as the Nationalists to take control.

As the fighting ground on, control of Stepan's town bounced between the Nationalists, and the sultan's supporters, and the Greeks. Eventually, many sought refuge in nearby towns, but Stepan remained behind. His stay was short-lived. In 1921, Adabazar changed hands again, falling under control of rebel Turks. As the Greeks withdrew, my grandfather and many other Christians followed, leaving their homes forever. Racing ahead of the Nationalists, Stepan and his family sprinted over the mountains, having left all their possessions behind. "Again, barefooted, we escaped, running, on an uneven road with rocks . . . all the way to Izmid, running, not walking," Ovsanna remembered. At some point, Stepan's brother-in-law died from an illness, leaving behind a widow and three girls. In Izmid, my grandfather and the others boarded a boat to Constantinople, where they moved into a cramped house shared by five refugee families.

With the capture of Adabazar, Nationalists killed nearly 80 percent of the six hundred Armenians remaining in town. For many Armenians, the genocide didn't end with the Treaty of Versailles. Kemal's forces continued to triumph, soon seizing Smyrna, which was set afire, sending Greeks and Armenians rushing toward the quay to await rescue by Allied ships. Tens of thousands more Armenians died during these tumultuous years after World War I, as did countless Greeks. The Assyrians also suffered terribly during the genocide with estimates of 275,000 casualties. The exact number of Armenians who lost their lives will forever be debated, but it has been grossly downplayed by the Turks to several hundred thousand. Stepan believed the toll was much higher than one million. He'd seen so many die before his eyes, he thought the count had to be double that figure.

The victory over the Greeks in the War of Independence paved the way for Mustafa Kemal's dream. Following the signing of the Treaty of Lausanne on July 24, 1923, Turkey became a new country and Mustafa Kemal its first president. The country then underwent a population exchange: Turkey's Greek inhabitants were traded for Greece's Turkish ones, with the country homogenizing. To Turks, Kemal would be forever known simply as Atatürk — "Father of the Turks" — and he is still revered for creating the modern secular state out of the rubble of the Ottoman Empire as the memory of the diverse people who once lived there was scrubbed away.

At the genocide memorial in Montebello, a California publisher, Harut Sassounian, strode before us and addressed the news many had already heard: President Barack Obama hadn't referred to the killings as *genocide* in his annual condolence message to the Armenians. Yet again, he'd used the words *Medz Yeghern,* Armenian for "Great Crime." I could feel the crowd's disappointment; I could see it in their creased faces. Sometimes, it felt like no one was listening to us, that this fight against powerful Turkey was futile and too painful to keep reliving year after year.

Harut, though, didn't waver. "President Obama has lost the moral standing," he said. "How can the United States leaders give a lecture to anybody around the world about justice, about democracy, about fairness, about peace, about preventing ethnic cleansing or genocide if they themselves are the top genocide denialist?"

The rapt crowd shook their heads and applauded. Many of those around me had immigrated to this country, their accents telling the story of their refugee families' postwar journey as they waited for accountability for the crimes.

While the British had lost faith in the Turkish courts, and transferred sixty-eight prisoners into their own detention for future prosecution, nothing ever came of it. Instead, there was a prisoner swap: British prisoners of war for Ottoman ones. The accused killers of the Armenians were released, most of them forever left unpunished for their crimes. Most, but not all. While strolling through Berlin one day, Talaat Pasha, one of the main architects of the genocide, felt a tap on his shoulder. When he turned around, a man shot him in the head, and he crumpled to the ground, dead. The gunman, Soghomon Tehlirian, had lost his entire family to the massacres, and believed his dead mother had directed him to avenge her murder. After listening to his testimony, a German jury acquitted him. The assassination had been orchestrated by the Armenian Tashnag Party's Operation Nemesis, which also oversaw the execution of a handful of other top CUP officers, including Jemal Pasha, the former minister of the navy.

In this new country of Turkey, my grandfather opened up a closet-size store with his brother, Armenag, selling hosiery, ribbons, and buttons. In 1925, when Stepan was thirty-eight and still single — a veritable crime! — his mother introduced him to the daughter of their neighbor Hagop Semizian, once a wealthy Adabazartsi merchant and now so financially ruined from the deportation he couldn't care for his youngest children, entrusting them to nuns instead. When Stepan met the lively and attractive Arshaluys Semizian, a fierce brunette fourteen years younger than him and an inch and a half taller, he was completely smitten. Two weeks later, they were married, and they soon had two girls, first Alice, and then my mother, Anahid. But his family was splintering. In September of 1921, his three younger sisters had left for New York. His mother followed, leaving Constantinople in 1927, and soon his sister Zaruhi departed for France with her family, including her daughter, Ovsanna. My grandmother Arshaluys was devastated when her own family immigrated to France.

With Turkish quotas met, Stepan couldn't enter the United States. Desperate, he paid a crooked priest to draw up papers stating that he and Arshaluys were Russian-born. It worked. Soon, the family of four boarded a ship and sailed away from Turkey forever. In the official documents, Stepan gave his birth year as 1892, making him six years younger than he actually was. He and Arshaluys had lied to appear more attractive to American employers. The truth was less promising: At forty-four years old, he was starting over again in America, with only a few dollars in his pocket and no English. After saying goodbye to Armenag, who would be the only one to remain behind, Stepan and his family went to France to visit relatives and then sailed across the rough Atlantic aboard the hulking SS *Statendam*, arriving in New York one week later, seasick and happy, on October 4, 1930. Finally, Stepan was free. He was on new soil, the kind where he could grow a life, the nightmare behind him.

In the months leading up to the genocide anniversary, my mother and I had been spending more time together. After many years of jokes about a *khosg gab*, I was finally engaged to a wonderful man, an *odar* named Steve, and our wedding loomed; we would be married five weeks after the 2012 march. My mother and I marveled at the coincidence, of how my name and Steve's in Armenian were exactly the same as my grandparents' (Arshaluys and Stepan); our marriage seemed predestined. With the wedding almost upon us, I was a little frantic, since I still didn't have a dress I liked. My mother, along with a coterie of friends, patiently accompanied me as I squeezed into gowns. My mother, ever practical, couldn't understand the fuss, telling us that she had bought her own A-line dress in downtown LA in one hour. But on her own wedding day, December 7, 1963, her father almost didn't attend. As the ceremony was poised to begin, she couldn't find him anywhere. At the last moment, he entered the chapel, reluctantly. A picture captures my grandfather Stepan near the elated couple, dressed in a brown suit, a white boutonniere pinned to his lapel. His lips curve upward in a sweet smile, but his eyes stare out, dark and heavy from pain. "It's too hard for me to be at joyful occasions," he explained to her later.

Yet he was living the American dream, even playing "God Bless America" regularly on his accordion. By stacking boxes in a grocery store in Los Angeles, where they'd moved during World War II, he and my grandmother, a factory seamstress, had managed to scrimp and save enough to invest in several apartment buildings as well as buy their own house. The couple had come a long way since they'd arrived in the United States with nothing and settled in Spanish Harlem in New York, where they soon had a son, John. With a one-thousand-dollar loan from Stepan's sister Mari, they had opened up a candy shop on West 133rd Street. Their Los Angeles life was a long way from my grandfather selling treats eighteen hours a day, seven days a week, just to have food and shelter, and my grandmother toiling in the garment district sewing endless nightgowns.

In that candy store, with huge placards for ice cream, cigarettes, and soda pop in the window, my grandfather often told his children about the massacres, the dark topic told against the backdrop of bub-blegum pink. Though out of danger, my grandfather would often warn my mother: "Anahid, I'm going to die next year." "Baba, don't say that," she'd implore. "But it's going to happen," he'd press. "No, Baba." Decades later, my mother would tell me, "Do you know what that was like, to think that your father is going to die each year? It seemed as if he could never reconcile with the fact that he had survived."

During the long drives to wedding-dress appointments, my mother and I would reflect on Baba's life. Then, with a heavy heart, we'd transition to the war that had broken out in Syria the year before, during the Arab Spring. With the rising death toll, history seemed to be repeating itself along the Euphrates. What had appeared at first to be a quick overthrow of President Bashar al-Assad had dragged out into a bloody civil war. Each day, my mother and I would worry all over again about the sheikh's family. I'd grown even closer to the sheikh's descendants after visiting again in 2009, and my mother and I considered them part of our own dispersed family across the globe. But now, in villages I'd visited, food was scarce, the price of bread too high, corpses commonplace. As one of the sheikh's descendants put it to me in a phone conversation, "We now know what your grandfather went through."

It would only get worse, it turned out. As the conflict escalated and the years went by, the so-called Islamic State of Iraq and Syria, or ISIS, proclaimed a "caliphate" in this region in 2014, enforcing a strict interpretation of Islamic Law, or sharia. Raqqa soon became its capital, and life for the civilians grew even more onerous. The city—where I had met the Bedouin sheikh who led me to Sheikh al-Aekleh's clan, and dined at a restaurant on the Euphrates alongside local Armenians and others of different beliefs — had been transformed. The new Raqqa had no room for religious unity, or tolerance. Jihadists beheaded foreigners, jettisoned gay men off rooftops, crucified Christians, and burnt Muslims alive in front of giant crowds while the cameras rolled.

They raised their black flag over the desert where my grandfather had walked, and where so many Armenians perished. Though no one has claimed responsibility, the genocide memorial in Deir Zor with the interred remains was blown up in 2014. The chapel's limestone was reduced to a pile of gray rubble, like so much of Syria. In Northern Iraq, jihadists massacred the Yazidis, a Kurdish-speaking minority, and forced the women into sexual slavery. In the region around Raqqa, they slaughtered other Muslims declared to be not faithful enough, and issued a death sentence to a kind man who had helped me during my second trip to Syria, requisitioning the home that he had spent his life's savings to buy and furnish. He fled to another part of Syria, displaced, much like the 8.7 million others, according to estimates by the United Nations High Commissioner for Refugees. Another 4.8 million became refugees, seeking safety outside the country's borders.

So many of the images in the news today are eerily similar to what my grandfather witnessed a century ago, long lines of people fleeing their homes and being marched to their deaths against a stark and unforgiving desert, adding to yet more mass graves, newly dug. All of these groups are being dehumanized and degraded as "other," just like countless people have been since the Armenians, whether Christian, Jewish, Muslim, or of another creed. A United Nations Human Rights panel called the killing in the region what it is: a genocide.

With war and persecution triggering the worst displacement on record, and an estimated 65.3 million pushed from their homes, so many are risking everything to reach safety. One was Omar, Sheikh al-Aekleh's relative. Setting out alone at night into the Syrian desert, with just a bag on his back, he eventually made it to Turkey. There, he waited with one hundred refugees to cross the Aegean Sea. He squeezed into one boat at double its capacity, while the other half of the group boarded a second craft. "You could not even move a hand," he said, as the waves lashed at them for nine hours, until reaching the coastline of Greece. He learned everyone else in the second vessel drowned: children, adults, the old and the young. Pressing on, he walked for weeks from country to country, through rain and cold, in a caravan of sometimes two hundred until his legs went numb. On the journey, partly by train, he rarely rested, subsisting on handouts of sardines, bread, and water. He is a young strong man, like my grandfather, and whenever he could, he tried to assist those struggling around him: the families with small kids, the elderly whom he supported. He did this because of his Muslim faith, he said, and what it teaches him about helping those in need. At other points, he trudged for miles without any food at all, until finally reaching his destination. "The next day, the feeling all over my body was pain," he recalled. "I couldn't move. Not just me, everyone."

Separated from his loved ones, and presently a refugee in Europe, he is trying to establish himself and to learn a new language, just like my grandfather once did. While welcomed at first, he now, too, feels the anti-Muslim sentiment that has soared in the wake of the Islamic State's worldwide terrorism campaign. He feels it in people's stares when he enters a business, or hears it in the whispers of passersby, but remains grateful to the kindest who see him and are not afraid, who see him as simply human. He wants to tell them all that he, too, is afraid of terrorism, committed by ISIS but also by President Assad, through what he has done to his own people. Omar has seen firsthand what all this violence has done to his village and the life that he and his family once had.

Others making the journey to leave Syria include the long established Armenian community — an estimated one hundred thousand before the war. Many are descendants of genocide survivors, and are now in search of a new home once more. Since the conflict began, hundreds of thousands have died, both Muslims and Christians alike. As the refugees pour into our communities, I think constantly of Sheikh Hammud al-Aekleh's core principle: to judge each person individually and to not give into fear about an entire group, which happened during my grandfather's era, and again during the Holocaust, and countless other times, but never should again.

Without any official accountability for the World War I crimes against the Armenians, some people grew tired of waiting and took justice into their own hands. I was ten years old on January 28, 1982, when two Armenians gunned down the Turkish consul general to Los Angeles as he sat in his car. I remember clearly watching my mother when she heard the news; her face whitened, and her jaw dropped. She was so saddened by the bloodshed. The consul general's death was part of an operation that assassinated Turkish emissaries around the world. My mother understood the motive — to revive awareness of the genocide and force Turkey to face its past — but not its method.

In recent years, though, there has been some progress, without such drastic actions. Article 301 of the Turkish Penal Code, which had outlawed insulting "Turkishness," was amended in 2008. While the language was softened, it continues to drastically limit freedom of expression, and remains a threat to those who might be tried under it. Still, in 2009, two hundred Turkish intellectuals bravely launched the "I Apologize" campaign online, which quickly gathered tens of thousands of signatures to the proclamation "My conscience does not accept the insensitivity showed to and the denial of the Great Catastrophe that the Armenians were subjected to in 1915." The Turkish government denounced the campaign, but the movement set the stage for the largest gesture, two years after that Montebello commemoration ceremony. In 2014, just before the ninety-ninth anniversary of the killings, the then Turkish prime minister, Recep Tayyip Erdoghan, expressed his sorrow over the "inhumane consequences" of the relocation and made the closest statement yet to an apology. "We wish that the Armenians who lost their lives in the context of the early twentieth century rest in peace, and we convey our condolences to their grandchildren," read his message. But still, he hadn't used the word *genocide*, and the momentum toward reconciliation was lost in 2015 when Erdoghan seemed to backtrack and adopt a harder stance. The centennial, though, saw Pope

Francis call the killings genocide, and Armenians increasingly begin to seek financial reparations from Turkey instead of official recognition.

Then, in June of 2016, acknowledgment came from a different source: Germany, the Ottoman Empire's former wartime ally. After Parliament voted to recognize the killings as genocide, Turkey blocked access to the Incirlik Air Base, leveraging its strategic position in the war in the Middle East, the influx of refugees into Europe, and Turkey's vital role in stemming the flow. Any momentum for genocide recognition within Turkey might have slowed that July with the failed military coup of Erdoghan, now president, in which an estimated 265 people were killed, as his government further cracked down all forms of dissent following the uprising, and those critical of the government.

Of the countless times my grandfather discussed the slaughter, my mother always circles back to something he said one particular afternoon. Her father had returned from a typical day of unclogging drains, fixing broken doors, and dealing with tenants in his apartments or another Adabazartsi's. He was always busy helping fellow immigrants, whether it was climbing onto a rooftop, at age eighty, to patch a friend's leak, carting supplies in a wheelbarrow for four miles across Hollywood, or feeding and sheltering the newly arrived in his own home. But on the day my mother remembers, he'd settled into his favorite burgundy armchair, lit a cigarette, and picked up the thread of a massacre story she knew well. This time, for some reason, my mother mustered the courage to ask the question that had burned inside her for thirty-some years, "Baba, should we forgive the Turks?"

He sat there for a long time in his armchair, his eyes drifting somewhere else. She looked at him, expectant, but he just sat there wordlessly. "He never said anything," she recounted. "It was a sad thoughtfulness that he couldn't express. But I take heart that he never said, 'No.' Dawn, he never said, 'No.'"

She believes he was thinking about the Turkish soldier who had saved his life and trying to find a space in his heart to forgive the killers. In his long silence, my mother found her own place to heal.

After a final prayer in Montebello, volunteers passed out carnations, and everyone queued to lay flowers at the monument. Holding ours, my mother and I stood at the end of the line that wrapped around the steps until we decided we'd pay our respects another way. We drove to the Hollywood Forever Cemetery, home to film legends like Jayne Mansfield and Rudolph Valentino — and to my grandfather, who died from natural causes in June of 1974 at the age of eighty-eight, four years after losing his wife.

I followed my mother down the winding road, both of us lost. "Here it is!" she yelled and made a beeline toward Stepan and Arshaluys's simple joint marker.

Standing over my grandparents' grave, I asked my mother what she wanted to say to them. Without glancing up, she switched to Armenian, "Mama and Baba, my daughter and I are together, we pray the day will come where there will be total understanding and forgiveness of what has happened." She turned to me, smiled, and placed her carnation atop the grave. I laid mine down and wished for the same.

ACKNOWLEDGMENTS

AT THE BEGINNING OF THIS LONG ROAD TO PUBLICATION bravely stands my grandfather, Stepan Miskjian. Without his courage to survive and his commitment to documenting these crimes against humanity, there would be no book, and there would be no me. I'd like to thank his three children for trusting me with their father's precious account — my late aunt Alice, my late uncle Johnny, and my mother, Anahid, who urged me to undertake the telling of his story. All of them inherited their father's remarkable spirit and strength of character and continue to inspire me each day.

What began as a two-year project turned into a decade-long one, as I set out to research primary materials in six foreign languages: Armenian, Ottoman Turkish, modern Turkish, Arabic, German, and French. It also took the patience of an extraordinary publisher, Houghton Mifflin Harcourt, whose extraordinary editor in chief at the time, Andrea Schultz, teamed me up with the brilliant and funny Nicole Angeloro. Nicole's vision allowed me to bring Stepan to life in these pages. She is the best champion a writer could have, and I already miss checking in with her every day. I'm also grateful to my acquiring editor, Webster Younce, and my subsequent editor, Anjali Singh, who both guided me as I began. And I'll never forget the support of my agent, David McCormick, as I took a leap of faith, quit my job, and moved back to Los Angeles. From the first moment, he was enthusiastic and supportive and he never wavered. With him in my corner, I remembered not to doubt myself.

After the genocide, survivors like my grandfather dispersed around the globe, along with their accounts. Following their trail took the aid of an army.

When I was stumbling in my research — without the language skills — fate brought me into the Ararat-Eskijian Museum in Mission Hills, California, and into the acquaintance of the lovely Maggie Mangassarian Goshin. I couldn't have done this book without her. From the beginning, she helped light my way around the globe, directing me to all the available archives, library collections, and people who could help expand my grandfather's story, such as the monks at the Mechitarist Monastery in Vienna. I'm grateful to them and their predecessors for safeguarding our history with care and compassion. I couldn't have found my way through their treasured collection without the aid of Father Simon Bayan, Father Vahan Hovagimian, Abbot Paul Kodjanian, and the indomitable assistant Anna Varosyan, who spent weeks hunched over the yellowed newspapers searching for articles about Adabazar. In addition to Anna, Madeleine Karacasian did invaluable research in Romania, Sarkis Zhamkochyan in Armenia, and Caner Dogan in Turkey. Maggie also connected me to my fixer, Baykar, who was the ideal guide for crossing Anatolia, always kind, resourceful, and unflappable.

Through his beautiful translations, the late novelist Agop Hacikyan gave my grandfather a voice in English, perfectly capturing his tone, and became a good friend over the years. John Kaldahar also translated part of my grandfather's memoirs and Adabazar-related material and then jumped in full force to complete the book's fact-checking. I can't thank both of them enough.

Sylvia Hakopian was the best research assistant I could ever have wished for, serving as my eyes to read the Armenian that I couldn't. She was as passionate as I was in following the clues about my grandfather's life, sharing in my joy when we discovered new details.

I couldn't have navigated my way through this complex subject without the numerous scholars who graciously answered question after question. From our first meeting in Aleppo, Raymond Kévorkian has always been patient and kind. During my visit to the AGBU's Nubarian Library in Paris, he pointed me to indispensable resources about the genocide and Adabazar. Richard Hovannisian provided access to his remarkable oral histories at UCLA and helped show me the way forward. Jay Winter at Yale University shored up my recounting of military events and patiently corrected my errors during his manuscript review. Vahakn Dadrian spent days with me at his New York home, and his generosity in sharing a lifetime of research gave me a deeper understanding of the subject matter. The historian Stephan Astourian at UC Berkeley was truly a lifesaver. He reviewed the book prior to publication, correcting mistakes and misinterpretations, and unifying my transliterations.

Yevkine LoMonaco shepherded my mother and me through my grandfather's journals. The late Arlene Yeran assisted with translations and we miss her all the time. Some of my warmest memories are at the Hollywood apartment of the late Knar Chalikian who fed me *lahmajun* while telling me about her mother, one of my grandfather's closest friends, and Adabazar. I'd like to thank my mother's dear friends who lent a hand with translation: Alice Balian, Shakeh Kargodorian, Stepan Karkodorian, Aghavni, and all the others who volunteered their time.

Alidz Agbabian and Armen Aroyan introduced me to Jemal, my caring and supportive driver through Turkey. Alidz also connected me with her beautiful aunt in Aleppo, who nursed me back to health when I was sick. While in Syria, Antranig was an indomitable driver, as he steered me across the country, along with my incredible fixer who introduced me to the late Sheikh Fayez al-Ghubein in Raqqa. Sheikh al-Ghubein, the embodiment of grace, helped me realize my dream of finding Sheikh Hammud al-Aekleh's descendants, and I'll always be in his debt. I'd also like to thank Mihran, my driver during my 2009 return trip to Syria, and my translator and his brother, who both helped translate Arabic into English.

The Hoover Institution, UCLA, the Krikor and Clara Zohrab Information Center, and the Ararat-Eskijian Museum all provided access to their collections. Verjiné Svazlian, Donald and Lorna Miller, Ani Shahinian, Garabet Moumdjian, Hagop Goudsouzian, Hilmar Kaiser, the late Harout Yeretzian, George Shirinian, Rouben Adalian, Aram Arkun, Sarkis Balmanoukian, Kevork Bardakjian, Sarah Freeman, Onnig Keshishian, and S. Peter Cowe all assisted at different stages. In addition, Dr. Peter Aronson at Yale patiently explained to me what happens to the body under extreme duress. Orlando Calumeno, Osman Köker,

Hayk Demoyan, Bardig Kouyoumdjian, Deutsche Bank, Ararat-Eskijian Museum, and Wallstein Verlag generously allowed me to republish photographs.

As I became consumed with my grandfather's story, I searched for those people he'd named in his journals. I won't forget meeting a descendant, Aida Kouyoumjian, after reading about how my grandfather sold water to her family in the Qatma camp. Aida shared the unpublished journals of her late mother, Mannig, with me, an account that was invaluable to understanding the conditions our families faced. Tracking down the relatives of my grandfather's friends and acquaintances also led me to Ara Chalvardjian and his niece Dr. Vania Chalvardjian in Canada. They told me about their family's flour mill near Tarsus, where my grandfather worked during the war. Armine Liberatore shared stories about her grandfather Tevon Harutunian and the family he left behind.

I also located my grandfather's friends by posting advertisements in newspapers. Osheen Keshishian helped me tremendously by placing mine in his publication, the *Armenian Observer*. This notice, which was similarly published in the *Armenian Reporter*, led me to Robert Roomian in Virginia, who shared a photograph of his grandfather Nerses Aghajanian, my grandfather's close friend. The posting also brought me into contact with many incredible Adabazartsis who told me all about their families, sharing photographs and books, all adding to the mosaic that was life in the town.

Writing about genocide was often grueling, and the following residencies provided me a peaceful sanctuary: Hedgebrook, Ragdale, Writers Omi at Ledig House, and the Edward F. Albee Foundation. The Taylor family hosted me in Michigan, the Brown family in Lake Tahoe, and Emilie Nakayama and Emil Dabora always gave me a quiet place to write in their home.

While trying to sift through my mountain of research documents, I could always turn to Amy Goldwasser. For years, she not only listened and guided me, but helped me construct the main arc of a story, editing draft after draft. Without her, I would have quit long ago. Following her lead was Vanessa Hua, who eagerly read pages and edited passages to smooth out the manuscript, staying up late into the night and motivating me onward, with intelligence and compassion. Janelle Brown was a gifted writing partner, lending her careful eye from the beginning to the manuscript's hectic completion, and she kept me from going crazy during our many writing retreats with her constant encouragement. Suzy Hansen not only helped me land in Istanbul in 2007 but provided an invaluable ear through the years about the complexities of the issue as well as a careful read of the manuscript. James Ellsworth was always ready to assist, spending days expertly polishing the story. I'm also eternally grateful to my other readers whose red-pen markups and keen insights indelibly shaped my pages: Mary Dittrich Orth, Tracy Roe, Benj Hewitt, Kerry Lauerman, Joan Walsh, Roni Rabin, Tom Clynes, Martha Kennedy, Keshni Kashyap, Lesley Alderman, Cristina Rueda, Ken Bensinger, Ruth Barrett, Kalee Thompson, Elizabeth Kairys Allspaw, Youssef Lemon, as well as Austen Wright and Emma Borges-Scott.

When personal challenges nearly stopped me from finishing this book, my family of friends cheered me on. Siubhan Lammas, Alyssa Hannah, and Simi Dhillon waved their pompoms until their arms tired, as did Jane Thomas, Anita Casanovas, Cynthia Joyce, Carina Chocano, Hala Youssef, Alicia Johnson, Dena Kobata, Yvonne Alvarado, Lori Leibovich, Kerri Higuchi, Vanessa Nilsson, Maria LaHood, Amy Marinelli, Sara Wood, Amanda Pike, Sara Wilson, Sue Jin Hwang, Patricia Yossen, and Laura Miller.

Writing a book about one's family means one gets to spend more time with them, and mine is an extraordinary one. Reconnecting with my mother's first cousin Ovsanna Kaloustian, who passed away in 2014 at age one hundred and six, was a precious gift. I am forever grateful to her, as she painstakingly recounted her experiences. Her daughter, Vartouhie, and grandchildren, Frederic and Florence, generously helped me reconstruct the whereabouts of my grandfather's family while he was fighting for his life. Ovsanna's beautiful spirit and resilience live on through her story and in these pages.

I cherished meeting my cousin Rehan in Istanbul for the first time. She assisted me in piecing together the life of the Miskjians and provided crucial assistance in translating Turkish. My cousin Carole Haspikian spent weeks translating French newspapers into English and helped me understand crucial genocide accounts. I'd like to thank cousins Jack and Mary Tartanian, Carol Neston, Rich and Marion Garry, Annette Hartunian, Sophie Haspikian, Richard Bremner, Catherine Ohanesian, Dorothy Davidian, Robert Davidian, and Stephanie Bridges.

My cousin, the editor and publisher Lois Bridges, never once doubted the importance of Baba's story and guided me forward when the way seemed murky. Her whole family contributed — Erin Bird, Brennan Bird, Gary Davidian, and especially her daughter, Dr. Aislinn Bird. My cousin, Jerry Miskjian, has also been extremely generous and supportive and I'm forever thankful to him, his wife, Aline, and his siblings, Stephan and Janine. My amazing uncles, John and Don MacKeen, and father-in-law, John, are the best surrogate fathers I could have. My sister-in-law Sandra, cousin Claudine LoMonaco, the Blatt family, and the magnificent Isabelle Bremner all encouraged me to keep moving forward, and I did.

But I truly stand each day, tall and strong, thanks to my husband, Steve. He is my best friend, my partner in adventure, and my reason to laugh. Through him, with each passing year, I learn even more about what it means to be a family.

I'm especially beholden to my late father, Jim MacKeen, who did everything possible to facilitate the writing of this book. Long ago, his paternal instinct kicked in and he wanted me to stop thinking about genocide every day, year after year. "Just finish the damn book," he'd say. Dad, I finally did, and I wish you could have read it.

Meeting Sheikh al-Aekleh's family has been the transcendent moment of my life. To Omar and his mother, Hala, I am so honored to have you be a part of my family and hope that in another hundred years our families will still be one.

Most important, I'd like to remember the million-plus Armenians who lost their lives during the genocide, as well as the ones who survived and somehow found the strength to continue. Nearly every Armenian family has a story like my grandfather's. This book is in their memory.

— Dawn Anahid MacKeen, November 2016

NOTES

A Note on Sources

Various contemporary geographical and cultural references not cited specifically were largely obtained from the following sources: Charles Edward Callwell, "Turkish Campaigns: Mesopotamian Operations," in *Encyclopaedia Britannica: The New Volumes*, 12th ed., vol. 17 (London: Encyclopaedia Britannica Company, 1922); Douglas Carruthers, "The Great Desert Caravan Route, Aleppo to Basra," *Geographical Journal* 52, no. 3 (September 1918); W. J. Childs, *Across Asia Minor on Foot* (Edinburgh: William Blackwood and Sons, 1917); W. J. Childs, "Turkey," *Encyclopaedia Britannica: The New Volumes*, 12th ed., vol. 17 (London: Encyclopaedia Britannica Company, 1922); Major-General Guy Payan Dawnay, "Turkish Campaigns: The Sinai Campaigns: 1916-1917," *The Encyclopaedia Britannica: The New Volumes*, 12th ed., vol. 17 (London: The Encyclopaedia Britannica Company, 1922); Major Franz Carl Endres, "Turkish Campaigns: Operations on the Caucasus Front," *The Encyclopaedia Britannica: The New Volumes*, 12th ed., vol. 17 (London: The Encyclopaedia Britannica Company, 1922); Christina Phelps Grant, *The Syrian Desert: Caravans, Travel and Exploration* (New York: Macmillan, 1938); Robert Francis Harper, "Down the Euphrates Valley I," *Old and New Testament Student* 10, no. 1 (January 1890); Robert Francis Harper, "Down the Euphrates Valley II," *Old and New Testament Student* 10, no. 2 (February 1890); Philip J. Haythornthwaite, *The World War One Source Book* (London: Arms and Armour Press, 1996); Ellsworth Huntington, "Railroads in Asia Minor," *Bulletin of the American Geographical Society* 41 (New York: American Geographical Society, 1909); C. G. Lewis, "A Survey of the Euphrates Valley from Deir-Ez-Zor to Aleppo," *Geographical Journal* 59, no. 6 (June 1922); P.H.H. Massy, "Exploration in Asiatic Turkey, 1896–1903," *Geographical Journal* 26 (July to December 1905); John Murray, *Handbook for Travellers in Constantinople, Brûsa, and the Troad* (London: John Murray, 1893); Alois Musil, *The Middle Euphrates: A Topographical Itinerary*, issue 3 (New York: American Geographical Society, 1927); Harry Pirie-Gordon, "Turkish Campaigns: The Palestine Campaign," *The Encyclopaedia Britannica: The New Volumes*, 12th ed., vol. 17 (London: The Encyclopaedia Britannica Company, 1922); William Warfield, *The Gate of Asia: A Journey from the Persian Gulf to the Black Sea* (New York: G. P. Putnam's Sons, 1916); "Euphrates," in *Encyclopaedia Britannica: A Dictionary of Arts, Sciences, Literature and General Information,* 11th ed., vol. 9 (New York: Encyclopaedia Britannica Company, 1910); "Mesopotamia," in *Encyclopedia Americana,* vol. 18 (New York: Encyclopedia Americana Corporation, 1919); "Physical Geography," Great Britain Naval Staff Intelligence Department, *C.B. 847A: A Handbook of Asia Minor: Volume 1, General* (July 1919).

Works cited frequently in the notes are referred to by the following abbreviations:

SM1   Stepan Miskjian memoirs, vol. 1, 1897–1913

SM2   Stepan Miskjian memoirs, vol. 2, 1902–1950

SM3   Stepan Miskjian memoirs, vol. 3, 1914–1915

SM4   Stepan Miskjian memoirs, vol. 4, 1915–1917

YDZ1   Stepan Miskjian, *Yes Der Zori Tzhokhke'n Pakhadz Em* [I Escaped from the Hell of Deir Zor], part 1: 1916 (Paris: Haratch, 1965)

YDZ2   Stepan Miskjian, *Yes Der Zori Tzhokhke'n Pakhadz Em* [I Escaped from the Hell of Deir Zor], part 2: 1916–1919 (Paris: Haratch, 1966)

RK   Raymond Kévorkian, *The Armenian Genocide: A Complete History* (London: I. B. Tauris, 2011)

Badmaqirk   Ardashes Biberian and Vartan Yeghisheyan, eds., *Badmaqirk Adabazar "Asdvadzareal" Kaghakin* [History Book of the "God-Created" City of Adabazar] (Paris: Der Agopian, 1960)

Kaloustian interviews   Ovsanna Kaloustian, interviews with the author, Marseille, France, from July 2 to 24, 2006

CHFWW   *The Cambridge History of the First World War,* volume 1, Global War, ed. Jay Winter (Cambridge: Cambridge University Press, 2014)

Official Records   Ara Sarafian, ed., *United States Official Records on the Armenian Genocide, 1915–1917* (Reading, England: Taderon, 2004)

"Armenian Genocide"   Wolfgang and Sigrid Gust, eds., "The Armenian Genocide 1915/16" (documents from the political archives of the German Foreign Office), http://www.armenocide.de/

The Lost World

6   *trains stuffed with people:* SM3, 15; YDZ1, 19, 23.

    demonized as a threat: General Directorate of State Archives, *Armenians in Ottoman*

Documents, 1915–1920 (Ankara: Başbakanlık Devlet Arşivleri Genel Müdürlüğü, 1995), 6–8.

two million Ottoman Armenians: RK, 272–78.

forced from their homes: James Viscount Bryce and Arthur J. Toynbee, eds., *The Treatment of Armenians in the Ottoman Empire* (1916; reprint, Beirut: G. Doniguian and Sons, 1989), xxi–xxiv.

1,200,000 Armenians perished: Ara Sarafian, *Talaat Pasha's Report on the Armenian Genocide, 1917* (London: Gomidas Institute, 2011), 6.

"Kill without pity": "Partial Text of Talks on Poland," *New York Times,* November 24, 1945.

7 *"murder of a nation":* Henry Morgenthau, *Ambassador Morgenthau's Story* (Garden City, NY: Doubleday, Page, 1919), 301.

twenty countries: "Chilean Senate Unanimously Recognizes Armenian Genocide," Armenian National Committee of America, press release, June 7, 2007.

Raphael Lemkin: Samantha Power, *A Problem from Hell: America and the Age of Genocide* (New York: Basic Books, 2002), 17–43.

"Being a witness": YDZ1, 18.

12 *"How I Became a Courier":* SM1, 15–36.

"An Anecdote from Pera, Constantinople": Ibid., 115–19.

"Return from the Armash Market": SM2, 37–42.

13 *now Adapazari:* John H. Coles and Howard E. Walker, *Postal Cancellations of the Ottoman Empire 4* (London: Christie's Robson Lowe, 1995), 9.

15 *"'Like pickled sardines'":* SM3, 19.

Empty Plans

18 *edged down the tracks:* SM1, 15–39.

this journey: "Chemin de Fer Ottoman D'Anatolie," *Stamboul,* September 30, 1914.

cluster of red-tiled rooftops: Osman Köker, ed., *Armenians in Turkey 100 Years Ago: With the Postcards from the Collection of Orlando Carlo Calumeno* (Istanbul: Birzamanlar Yayıncılık, 2005), 104–6.

19 *Haydarpasha:* Edward Mead Earle, *Turkey, the Great Powers, and the Bagdad Railway: A Study in Imperialism* (New York: Macmillan, 1924), 86.

connecting the rail: Ibid., 21, 30–31, 36.

20 *"Aladdin and His Magic Lamp":* Robert Hichens, *The Near East: Dalmatia, Greece, and Constantinople* (New York: De Vinne Press, 1913), 192.

darkened hamams: Ibid., 201.

Christian past: Philip Mansel, *Constantinople: City of the World's Desire, 1453–1924* (London: John Murray, 2006), 1–2.

tombs: Hichens, *The Near East,* 201.

Mahmud Pasha Street: Köker, *Armenians in Turkey,* 76.

father, Hovhannes, had died: SM1, 11; YDZ1, 81.

married and moved out: Kaloustian interviews.

good name alone: SM1, 12–13.

21 *three-tiered house:* Kaloustian interviews.

widowed mother slept: SM1, 49.

dressed in black: This is according to Anahid Miskjian's account of her grandmother Hripsime as well as family photographs.

fine slipcovers: SM2, 44–48.

Nemcheler Street: YDZ1, 38.

arched its back: Vartan Yeghisheyan, map of Adabazar, in *Asdvadzareal Kaghak* (Paris: Der Agopian, 1937), preface.

small flower boxes: Köker, *Armenians in Turkey*, 112.

of wood and stone: Badmaqirk, 421.

wooded hills: S. M. Dzotsigian, *Arevmdahay Ashkhar* [Western Armenian World] (New York: Leylegian, 1947), 85.

The town slowly awakened: Köker, *Armenians in Turkey*, 112–13.

silkworms for extra cash: Jack Tartanian, Arshaluys's son, interview with the author, summer 2006.

empire of some twenty million: Edward J. Erickson, *Ordered to Die: A History of the Ottoman Army in the First World War* (Westport, CT: Greenwood, 2001), 16.

thirty thousand residents: Badmaqirk, 87; *Annuaire Oriental, 1914: Commerce, Industrie Administration, Magistrature de L'Orient* (Constantinople: Autrrichienne Ferd. Walla, 1914) 88. Population estimates range from 25,000 to 40,000.

22 *formally accept Christianity:* Louise Nalbandian, *The Armenian Revolutionary Movement* (Berkeley: University of California Press, 1963), 8.

The four quarters: Badmaqirk, 90.

named after the churches: Ibid., 91.

shots of oghi: SM2, 17–18.

God-Created City: Dzotsigian, *Arevmdahay Ashkhar*, 82.

the intermittent plagues: Badmaqirk, 420–23.

four inches above Stepan: Relative who wishes to remain anonymous, interview with the author, August 29, 2014.

23 *within his sanjak:* Badmaqirk, 411.

blazed into power: M. Şükrü Hanioğlu, *Preparation for a Revolution: The Young Turks, 1902–1908* (New York: Oxford University Press, 2001), 266–78.

switched sides: Ibid., 278.

celebrations of their victory: Mansel, *Constantinople*, 346.

clogged the roads: Badmaqirk, 409–11.

liberty, equality, fraternity: Mechveret: *Organe de la Jeune Turquie, Supplément Français*, August 1, 1908, 2.

millets: Taner Akçam, *A Shameful Act: The Armenian Genocide and the Question of Turkish Responsibility* (New York: Metropolitan Books, 2006), 23–25; RK, 23–24, 71.

treated them unfairly: Vahakn N. Dadrian, *The History of the Armenian Genocide: Ethnic Conflict from the Balkans to Anatolia to the Caucasus* (New York: Berghahn Books, 2004), 114.

fewer rights: Akçam, *A Shameful Act*, 36–37.

24 *Armenian kings*: Nalbandian, *The Armenian Revolutionary Movement*, 2–3.

ancestral homeland: Ibid.

thousands of years: Stephan H. Astourian, "In Search of Their Forefathers: National Identity and the Historiography and Politics of Armenian, an Azerbaijani Ethnogeneses," in *Nationalism and History: The Politics of Nation Building in Post-Soviet Armenia, Azerbaijan and Georgia*, ed. Donald V. Schwartz and Razmik Panossian (Toronto: University of Toronto, 1994), 46.

the Persians, the Macedonians: Nalbandian, *The Armenian Revolutionary Movement*, 4–5.

the Seljuk Turks: Ibid., 12–14.

They wanted reforms: Akçam, *A Shameful Act*, 35–39.

a new constitution: Ibid., 31.

was quickly suspended: RK, 9.

Pockets of Armenians rose up: Akçam, *A Shameful Act*, 41.

two hundred thousand Armenians were massacred: Dadrian, *The History of the Armenian Genocide*, 155–56.

"the Red Sultan": Stephen Bonsal, "'The Red Sultan' — As Seen in the Yildiz Kiosque," *New York Times*, April 25, 1909.

sultan was exiled: Peter Balakian, *The Burning Tigris* (New York: HarperCollins, 2004), 152.

a provisional armed unit: Badmaqirk, 415–18.

25 *Island Bazaar*: Ibid., 55.

unprecedented cultural awakening: Ibid., 411.

two hundred books: Yenovk Armen, "Out of Constantinople," *Piût'ania* 1 (January 1, 1910): 2–3.

"The true readers": Ibid.

credit unions: Badmaqirk, 411.

caterpillar-size mustache: Hagop F. Malkhasi, "Instananeous Coloring: Hair, Mustache and Beard," *Piût'ania* 6 (February 20, 1910): 60.

26 *yassas*: Anahid MacKeen's memories of her father.

27 *eleven times what he used to earn*: SM1, 12–13.

Hill of Liberty: SM2, 94–97.

buried together in one grave: Sir Edwin Pears, *Forty Years in Constantinople* (New York: D. Appleton, 1916), 282.

perfectly straight: Anahid MacKeen's description of her father's joke-telling.

one hundred gold pounds: SM1, 27–29.

28 *women's rights*: "Who Will Defend the Rights of a Woman," *Piût'ania* 13 (May 20, 1911): 425–26.

a peep to her father-in-law: Dzotsigian, *Arevmdahay Ashkhar*, 83.

because they finally could: Vahan Chavushian, "Societies [in Adabazar]," in *Badmaqirk*, 276.

silkworm mills: Serope Burmaian, "Union of Armenian Friends," in *Badmaqirk*, 263.

Armenian citizens to assemble: Chavushian, "Societies [in Adabazar]," 275–77.

auditorium should be modern: SM1, 114.

two thousand people: Chavushian, "Societies [in Adabazar]," 276.

Two drunken men: "Attack on a Hat," *Piut'ania* (January 5, 1912).

new Armenian press: "Local," *Piut'ania* 23 (October 22, 1911): 501.

round glasses: Badmaqirk, 319.

Egypt had practically become British: David Fromkin, *A Peace to End All Peace* (New York: Henry Holt, 1989), 85.

the Dardanelles: "Keep Dardanelles Closed," *New York Times*, April 29, 1912.

29   *handful of Ottoman islands:* Richard C. Hall, *The Balkan Wars, 1912–1913: Prelude to the First World War* (New York: Routledge, 2000), 19–20.

"Sick Man of Europe": Akçam, *A Shameful Act*, 27.

Bulgaria had announced its autonomy: Ibid., 55.

positions in the cabinet: Fikret Adanır, "Non-Muslims in the Ottoman Army and the Ottoman Defeat in the Balkan War," in *A Question of Genocide: Armenians and Turks at the End of the Ottoman Empire,* ed. Ronald Grigor Suny, Fatma Müge Göçek, and Norman M. Naimark (New York: Oxford University Press, 2011), 118–19.

Italians ended: Hall, *The Balkan Wars*, 20.

that October: Ibid., 39.

drafted: SM1, 41–42.

Adrianople: Carnegie Endowment for International Peace, *Report of the International Commission to Inquire into the Causes and Conduct of the Balkan Wars* (Washington, DC: Carnegie Endowment, 1914), 110–16.

The Countdown

30   *empire's worst slaughterhouses:* YDZ1, 5, 119.

32   *blame the victims:* Turkish Republic Prime Ministry General Directorate of the State Archives, *Armenians in Ottoman Documents, 1915–1920* (Ankara: Başbakanlık Devlet Arşivleri Genel Müdürlüğü, 1995), 6–9.

33   *"A prominent newspaper editor":* Sednem Arsu, "Armenian Editor Is Slain in Turkey," *New York Times*, January 20, 2007.

a rumpled white sheet: Joel Roberts, "Turkish-Armenian Journalist Shot Dead," CBS News/AP, January 19, 2007.

The assassin: Associated Press, "Scandal Tails Death of Turkey Journalist," *Washington Post*, February 2, 2007.

The teen's supposed motive: Sebnem Arsu and Susanne Fowler, "Turkish Gunman Said to Confess to Killing Armenian Editor," *New York Times*, January 22, 2007.

"insulting Turkishness": Sabrina Tavernise, "Trial in Editor's Killing Opens, Testing Rule of Law in Turkey," *New York Times*, July 3, 2007.

Turkish penal code: Arsu, "Armenian Editor Is Slain."

"A million Armenians": Maureen Freely, "I Stand By My Words," *Guardian*, October 22, 2005.

"*Armenian question*": Turkish Republic Prime Ministry General Directorate of the State Archives, *Armenians in Ottoman Documents, 1915–1920* (Ankara: Başbakanlık Deviet Arşivleri Genel Müdürlxüğü, 1995), x, xi.

"*so-called Armenian genocide*": "Views Against Genocide Allegations," Ministry of Culture and Tourism, http://www.kultur.gov.tr/.

Alphabet City

35 *agonizing months*: SMI, 42–55, 77.

spinning wool: Kaloustian interviews.

starved: Ibid.

cold winter: Carnegie Endowment for International Peace, *Report of the International Commission to Inquire into the Causes and Conduct of the Balkan Wars* Publication No. 4 (Washington, DC: Carnegie Endowment, 1914), 113.

corpse of a horse: Kaloustian interviews.

stage a coup d'état: Lord Kinross, *The Ottoman Centuries: The Rise and Fall of the Turkish Empire* (New York: Morrow Quill, 1979), 595.

second short war: Richard C. Hall, *The Balkan Wars, 1912–1913: Prelude to the First World War* (New York: Routledge, 2000), 1.

tremendous losses: Donald Bloxham, *The Great Game of Genocide: Imperialism, Nationalism, and the Destruction of the Ottoman Armenians* (Oxford: Oxford University Press, 2009), 62–63.

most of their European territories: Taner Akçam, *A Shameful Act: The Armenian Genocide and the Question of Turkish Responsibility* (New York: Metropolitan Books, 2006), 8.

displaced Muslim families: Bloxham, *The Great Game of Genocide*, 63.

36 *ethnic and religious inclusiveness*: Fikret Adanır, "Non-Muslims in the Ottoman Army and the Ottoman Defeat in the Balkan War," in *A Question of Genocide: Armenians and Turks at the End of the Ottoman Empire*, ed. Ronald Grigor Suny, Fatma Müge Göçek, and Norman M. Naimark, (New York: Oxford University Press, 2011), 113, 124–25.

set the stage for more turmoil: Hall, *The Balkan Wars*, 129.

Armenag by his side: SMI, 77–80.

purchase land: Badmaqirk, 263–64, 277–79.

Sahag and Mesrop Mashdotz: S. Stepanian, "A Bloody Fifteenth Centenary: A Cultural Celebration Turns Bloody," in *Badmaqirk*, 436–37.

foundation of Armenian identity: "Le Grand Jubilé Arménien 1513–1913: La Fête de la Civilisation Arménienne," *Stamboul*, October 25, 1913.

"*Celebration of Armenian Civilization*": Ibid.

"*beautifully calm*": "Bulletin Meteorologique," *Stamboul*, October 27, 1913.

stained-glass: "Photo of Archbishop Stepan Hovagimian Visiting the Surp Hreshdagabed Church," in *Badmaqirk*, 352.

37 *modest attendance*: Yenovk Armen, "Out of Constantinople," *Piütania* I (January 1, 1910): 2–3.

triangular hood: Badmaqirk, 329.

the next two hours: Stepanian, "A Bloody Fifteenth Centenary," 436.

first to parade: Ibid., 434–47.

eighteen pairs of boys and girls: Ibid., 437.

six vilayets: Louise Nalbandian, *The Armenian Revolutionary Movement* (Berkeley: University of California Press, 1963), 25.

Russian-Armenians who had established: Gerard J. Libaridian, "What Was Revolutionary about Armenian Revolutionary Parties," in *A Question of Genocide,* 87.

both countries: Ibid., 102.

in Russia, the Armenians: "Regarding the Event in Adabazar," *Azadamard,* October 19/November 1, 1913, no. 1342.

38 *Tashnag Party had stopped supporting:* Hans-Lukas Kieser and Donald Bloxham, "Genocide," in CHFWW, 598.

"internal tumors": Akçam, *A Shameful Act,* 90–92.

pressed for reforms again: Ibid., 97–101.

similar celebrations: "On the Occasion of the Armenian Demonstration," *Puzantion,* October 15–28, 1913.

"carried their audacity": Djemal Pasha, *Memories of a Turkish Statesman, 1913–1919* (New York: George H. Doran, 1922), 265.

"The march is forbidden": Stepanian, "A Bloody Fifteenth Centenary," 438.

39 *grand finale:* Ibid., 437.

his lean body: Badmaqirk, 699.

"a peaceful procession": Stepanian, "A Bloody Fifteenth Centenary," 438.

loaded their weapons: "The Tragic Incident of Adabazar," *Puzantion,* October 15–28, 1913.

the drums tapped: Badmaqirk, 278.

"Fire": Stepanian, "A Bloody Fifteenth Centenary," 438.

At three o'clock: "The Tragic Incident of Adabazar."

40 *"I don't have my P":* Stepanian, "A Bloody Fifteenth Centenary," 439.

"Met with blows": "The Tragic Incident of Adabazar."

In the forty-eight hours that followed: "The Incident of Adabazar," *Puzantion,* October 16–29, 1913.

businesses were closed: "The Tragic Incident of Adabazar."

A curfew was imposed: "The Incident of Adabazar."

calm a nervous populace: "Incident in Adabazar," *Puzantion,* October 17–30, 1913.

one hundred and fifty Armenian homes searched: "The Incident in Adabazar," *Puzantion,* October 19–November 1, 1913.

Only licensed hunting rifles: Badmaqirk, 452.

posters of the Armenian political parties: "The Tragic Incident of Adabazar."

Dikran, Krikor, and many others were released: "The Incident of Dikran," *Puzantion,* October 24–November 6, 1913.

guilty of murder: "The Incident in Adabazar."

fallen Turkish soldier was laid to rest: "Soldier Buried," *Puzantion,* October 18–31, 1913.

"Don't forget to avenge me": Stepanian, "A Bloody Fifteenth Centenary," 447.

41 *gun dealer*: SM1, 98.

the trip to the capital: Ibid., 98–114.

42 *children's socks*: Ibid., 81–97.

"under the black tree": Badmaqirk, 458.

44 *stacks of textiles*: Osman Köker, ed., *Armenians in Turkey 100 Years Ago: With the Postcards from the Collection of Orlando Carlo Calumeno* (Istanbul: Birzamanlar Yayıncılık, 2005), 78.

carpets, and jewels: Robert Hichens, *The Near East: Dalmatia, Greece, and Constantinople* (New York: De Vinne Press, 1913), 202–5.

Blue Mosque: Carlton D. Harris, *Through Palestine with Tent and Donkey and Travels in Other Lands* (Baltimore: Southern Methodist Publishing, 1913), 95.

become a scapegoat: SM1, 109.

the Central Prison: Ibid.

45 *dating back hundreds of years*: Clarence Richard Johnson, ed., *Constantinople Today: The Pathfinder Survey of Constantinople* (New York: Macmillan, 1922), 336–38.

"101 years": Ibid., 326.

46 *the small chapel*: Ibid., 342.

Breaking Stones

49 *new assembly hall*: SM1, 114.

Theater! Music! Lectures: Serope Burmaian, "Union of Armenian Friends," in *Badmaqirk*, 263–64.

ten o'clock: SM3, 1–4, 11–12.

August: RK, 179.

"the ages of twenty and forty-five": Ibid. This corrects Stepan's account, which states "between eighteen and forty-five."

50 *Archduke Franz Ferdinand*: Charles Willis Thompson, "Anniversary of the War's Origin," *New York Times*, June 27, 1915.

Yugoslav nationalist: Jean-Jacques Becker and Gerd Krumeich, "1914: Outbreak," in *CHFWW*, 42–43.

the recent Balkan Wars: Ibid.; Volker R. Berghahn, "Origins," in *CHFWW*, 32–33.

primed the continent to blow: Berghahn, "Origins," 20–33.

secured Germany's support: Becker and Krumeich, "1914: Outbreak," 46.

waged war on Serbia: Ibid., 39.

Russia mobilized in Serbia's defense: Ibid., 53–54.

Europe marched into war: Thompson, "Anniversary of the War's Origin."

to be precautionary: "Turkey Mobilizes Also: But Informs Great Britain That She Will Remain Neutral," *New York Times*, August 5, 1914.

secret treaty with the Germans: RK, 178–79.

disastrous Balkan war: SM1, 41–54.

forty-two-gold-pound fee: Antranig Genjian, ed., *Haladzagani Husher, 1914–1918*

[A Fugitive's Memoirs] (Beirut: Ararat's Madenashar, 1964), 5. This corrects Stepan's citation of forty-four gold pounds; the forty-two figure is also given in RK, 388.

51 *in the summer heat:* Zaven Der Yeghiayan, *My Patriarchal Memoirs* (Barrington, RI: Mayreni, 2002), 34.

queues for bread: Philip Mansel, *Constantinople: City of the World's Desire, 1453–1924* (London: John Murray, 2006), 373.

procure provisions: "The Ottoman Maneuver," *Puzantion,* July 21–August 3, 1914.

"All soldiers are required": Ibid.

death penalty: Henry Morgenthau, *Ambassador Morgenthau's Story* (Garden City, NY: Doubleday, Page, 1919), 66.

hookah water: Yervant Odian, *Accursed Years: My Exile and Return from Der Zor, 1914–1919,* trans. Ara Stepan Melkonian (London: Gomidas Institute, 2009), 26.

the United States declared its neutrality: "President Wilson Proclaims Our Strict Neutrality; Bars All Aid to Belligerents and Defines the Law," *New York Times,* August 5, 1914.

"Paris for lunch, dinner in St. Petersburg": J. M. Winter and Blaine Baggett, *The Great War and the Shaping of the Twentieth Century* (New York: Penguin Studio, 1996), 59.

this time with Bulgaria: Edward J. Erickson, *Ordered to Die: A History of the Ottoman Army in the First World War* (Westport, CT: Greenwood, 2001), 37.

52 *labor battalion:* SM3, 4–9.

Along with the Greeks: Taner Akçam, *A Shameful Act: The Armenian Genocide and the Question of Turkish Responsibility* (New York: Metropolitan Books, 2006), 142.

pack animals: Morgenthau, *Ambassador Morgenthau's Story,* 302.

plowed to a halt: RK, 179.

53 *heart-shaped face:* Photograph supplied to author by Armine Liberatore, grand-daughter of Tevon Harutunian, October 2014.

high fever: Armine Liberatore, interview with the author, Santa Clarita, California, June 29, 2007.

monitor Armenian leaders: Akçam, *A Shameful Act,* 141.

enough food: "Bread Prices," *Puzantion,* July 21–August 4, 1914.

wheat for two months: "Military," *Puzantion,* July 25–August 7, 1914.

"Give us bread!": Mansel, *Constantinople,* 373.

melodies in Turkish or Armenian: Verjiné Svazlian, *The Armenian Genocide: Testimonies of the Eyewitness Survivors* (Yerevan: Gitut'yun Publishing, 2011), 19.

accordion: Anahid Miskjian MacKeen's memory of her father.

kiss their hands: Svazlian, *The Armenian Genocide,* 553.

leaving their beloved and dear brides behind: Ibid., 551.

"Mother, Mother": Ibid., 552.

54 *dug trenches across Europe:* David Fromkin, *A Peace to End All Peace* (New York: Henry Holt, 1989), 126.

attacked Russian seaports: Ibid., 71–73.

Russia declared war on the Ottoman Empire: Erickson, *Ordered to Die,* 35–36.

Great Britain and France: "Britain and France Have Declared War," *New York Times,* November 6, 1914.

repealed the capitulations: Fromkin, *A Peace to End All Peace,* 47.

establishments shuttered: Henry Morgenthau, *United States Diplomacy on the Bosphorus,* ed. Ara Sarafian (Reading, England: Taderon, 2004), 131.

"Death to Russia": Mansel, *Constantinople,* 373.

holy war: Liman von Sanders, *Five Years in Turkey* (Baltimore: Williams and Wilkins, 1928), 34–35; Robert M. Labaree, "The 'Jihad' Rampant in Persia," in *The Armenian Genocide: News Accounts from the American Press: 1915–1922,* ed. Richard Diran Kloian (Richmond, CA: Heritage, 2007), 17–19.

flooded the train station: Morgenthau, *Ambassador Morgenthau's Story,* 133–36.

negotiated a deal: SM3, 9–10.

55   *the young men away:* Sos-Vani, "The Deportation," in *Badmaqirk,* 702–3.

women and teenagers manned the shops: Ibid., 703.

conclude within six months: *Badmaqirk,* 369.

two weeks: Ibid.

travel prohibited: Genjian, *Haladzagani Husher,* 5–6.

Basra had been lost: Charles Townshend, *Desert Hell: The British Invasion of Mesopotamia* (Cambridge, MA: Belknap Press of Harvard University Press, 2011), 34–35.

Russia's eastern border: von Sanders, *Five Years in Turkey,* 38.

halting a Russian advance: Ibid., 37.

thirty degrees below: Fromkin, *A Peace to End All Peace,* 120.

56   *own ineptitude and failure:* von Sanders, *Five Years in Turkey,* 38–40.

Armenians did fight elsewhere: RK, 242.

revenge: Alan Moorehead, *Gallipoli* (New York: HarperCollins, 2002), 90.

"the propaganda work necessary": Akçam, *A Shameful Act,* 125.

top of the world's news: "Silence Forts on Dardanelles," *New York Times,* February 21, 1915.

armada of French and English warships: Moorehead, *Gallipoli,* 45–46.

Suez Canal: Fromkin, *A Peace to End All Peace,* 121.

take Constantinople: Moorehead, *Gallipoli,* 29–32.

hasten the war's end: Fromkin, *A Peace to End All Peace,* 128.

Gallipoli campaign battered on: Moorehead, *Gallipoli,* 70.

the archives, and the gold: Fromkin, *A Peace to End All Peace,* 152.

repel the naval assault: Moorehead, *Gallipoli,* 84–87.

57   *bolstered the Turkish leaders' resolve:* Ibid., 87–90.

Stepan found an unwanted surprise: SM3, 11.

"Armenian individuals are absolutely not": Akçam, *A Shameful Act,* 144.

"sealed letters": Sos-Vani, "The Deportation," 703.

"The Turks were silent": Ibid.

"In almost all cases": Morgenthau, *Ambassador Morgenthau's Story,* 302–3.

58   *weren't buried:* Germany, Turkey, and Armenia: A Selection of Documentary Evidence

Relating to the Armenian Atrocities from German and Other Sources (London: J. J. Keliher, 1917), 84.

That April: Genjian, *Haladzagani Husher,* 6. Note: The author gives the date as April 14, 1915, in the Julian calendar, which translates to late April 1915 in the Gregorian calendar.

Then the arrests rolled out: Ibid., 8. Genjian gives April 22, 1915, as the date the first caravan departed, which converts to May 1915.

socialist Hnchag political party: Gerard J. Libaridian, "What Was Revolutionary about Armenian Revolutionary Parties," in *A Question of Genocide: Armenians and Turks at the End of the Ottoman Empire,* ed. Ronald Grigor Suny, Fatma Müge Göçek, and Norman M. Naimark (New York: Oxford University Press, 2011), 84–85.

"We didn't know where": Genjian, *Haladzagani Husher,* 8.

April 24, 1915: Mikayel Shamtanchian, *The Fatal Night: An Eyewitness Account of the Extermination of Armenian Intellectuals in 1915,* trans. Ishkhan Jinbashian (Studio City, CA: H. and K. Manjikian, 2007), 1–2.

been asleep: Grigoris Balakian, *Armenian Golgotha,* trans. Peter Balakian and Aris Sevag (New York: Alfred A. Knopf, 2009), 56.

who's who of Armenians: Shamtanchian, *The Fatal Night,* 2–5.

59 *senators, deputies:* Mansel, *Constantinople,* 375.

"You, too": Shamtanchian, *The Fatal Night,* 9.

Scared and anxious: Balakian, *Armenian Golgotha,* 57–58.

a blacklist: Ibid., 58.

boated over to Haydarpasha: Ibid., 59–60.

trains took them eastward: Shamtanchian, *The Fatal Night,* 13–17.

never to be heard from again: Ibid., 58–66.

The arrested men of Adabazar soon vanished: Genjian, *Haladzagani Husher,* 8.

outside the town of Konya: Ibid., 10.

Hagop Semizian: YDZ2, 78.

men in Stepan's hometown agonized: Genjian, *Haladzagani Husher,* 11.

no place held a safe harbor anymore: SM3, 11–12.

People We Don't Mention

61 *practical jokes:* SM2, 89–91.

nearby road: SM1, 116.

62 *telegram his family:* SM1, 42–51.

Hovsep Aznavurian: "A Man Who Holds the History of the Famous Mısır Apartment," *Today's Zaman,* January 18, 2011.

derelict edifice: Vercihan Ziflioğlu, "Landmark Istanbul Hotel Threatened by Stall on Restoration," *Hürriyet Daily News,* December 27, 2010.

63 *thousands protested:* Sebnem Arsu and Susanne Fowler, "Turkish Gunman Said to Confess to Killing Armenian Editor," *New York Times,* January 22, 2007.

We Are All Hrant Dink: Yesim Borg and Laura King, "Armenians Bid Goodbye to a Hero," *Los Angeles Times,* January 24, 2007.

64 *"an unknown path":* SM1, 1.

Following Orders

65 *war-torn Dardanelles:* SM3, 11–12.

the upheaval: Ibid., 12–15.

the people of Sabanja: RK, 552; Sophie Holt, "Fuller Statement by the Author of the Preceding Document," in *The Treatment of Armenians in the Ottoman Empire,* ed. James Viscount Bryce and Arthur J. Toynbee (1916; reprint, Beirut: G. Doniguian and Sons, 1989), 402.

nearby town of Bardizag: Sebuh Aguni, *Milion Me Hayeru Charti Badmutiune* [History of the Massacre of a Million Armenians] (Constantinople: H. Asadourian and Sons, 1921), 275–76.

fifteen thousand troops: Ibid., 266.

66 *wide avenue:* Osman Köker, ed., *Armenians in Turkey 100 Years Ago: With the Postcards from the Collection of Orlando Carlo Calumeno* (Istanbul: Birzamanlar Yayincilik, 2005), 112.

operated as a hospital: Mihran Mavian, "The Deportation," in *Badmaqirk,* 710.

smallest of infractions: Aguni, *Milion Me Hayeru Charti Badmutiune,* 266.

Ibrahim Bey arrived: Sos-Vani, "The Deportation," in *Badmaqirk,* 705–7. Sos-Vani is the pen name of Garabed Hovhannesian.

reputation had preceded him: Ibid., 705. See also Aguni, *Milion Me Hayeru Charti Badmutiune,* 266.

Surp Garabed Church: Stepan Ishkhanian, "Transcribed from His Journals," in *Haladzagani Husher, 1914–1918* [A Fugitive's Memoirs], ed. Antranig Genjian (Beirut: Ararat's Madenashar, 1964), 130–42. Using the Julian calendar, Ishkhanian states the arrest began on July 7 and ended on July 16.

"Let me first introduce myself": Aguni, *Milion Me Hayeru Charti Badmutiune,* 267–69.

clergy: Sos-Vani, "The Deportation," 708.

67 *massacred some twenty-five thousand Armenians:* Vahakn N. Dadrian, *The History of the Armenian Genocide: Ethnic Conflict from the Balkans to Anatolia to the Caucasus* (New York: Berghahn Books, 2004), 181–82.

held on to the arms for protection: Sos-Vani, "The Deportation," 704.

"They aren't dissolved": Aguni, *Milion Me Hayeru Charti Badmutiune,* 268.

Mosul was a vilayet: Charles Townshend, *Desert Hell: The British Invasion of Mesopotamia* (Cambridge, MA: Belknap Press of Harvard University Press, 2011), 6.

"ten percent of the Muslim population": Turkish Republic Prime Ministry General Directorate of State Archives, "Expanding the Area Where Armenians Will Be Settled, and to Settle Them Based on a Ratio of 10% of the Muslim

Population," in *Armenians in Ottoman Documents, 1915–1920* (Ankara: Başbakanlık Devlet Arşivleri Genel Müdürlüğü, 1995), 67.

68 falaka: Sos-Vani, "The Deportation," 707–9.

knowing they'd be next: Ishkhanian, "Transcribed from His Journals," 132.

feet burst open: W. Hunecke, "Miss Hunecke's Report, Everek," in *Official Records*, 114.

Reverend Mikayel Yeramian: Sos-Vani, "The Deportation," 708.

screams were so loud: Arsha Louise Armaghanian, *Arsha's World and Yours* (New York: Vantage Press, 1977), 12.

for a warning: Ishkhanian, "Transcribed from His Journals," 133.

to six hundred: Aguni, *Milion Me Hayeru Charti Badmutiune*, 269.

"They are beating the men": Holt, "Fuller Statement by the Author of the Preceding Document," 401. Note: Holt's earlier account — "Adapazar: Statement, Dated 24th September, 1915, by a Foreign Resident in Turkey; Communicated by the American Committee for Armenian and Syrian Relief," 398 — reports that the beating began on August 1, but the author is following the diary of Ishkhanian, "Transcribed from His Journals," 127–43.

fear descended: Holt, "Fuller Statement," 401–2.

from their own gardens: Ibid., 402.

women kept vigil: Ibid., 400–401.

her disabled son: Ibid., 400.

"Get out of the way": Ibid.

the minister of the interior: Robin Prior, "The Ottoman Front," in *CHFWW*, 297.

69 *two carts rolled up*: Sos-Vani, "The Deportation," 708.

"no God but me": Holt, "Fuller Statement," 400.

marking off their names: Ibid., 401.

"Why do you punish these men": Ibid., 402.

hidden these weapons: Sos-Vani, "The Deportation," 703–4.

With a high forehead: *Badmaqirk*, 205.

Harutiun's betrayal: Sos-Vani, "The Deportation," 706–7.

his wife, Lusi: *YDZ1*, 88.

hot eggs: Sos-Vani, "The Deportation," 707.

wished for death: Ibid., 708.

A first cousin of Stepan's: Ibid.

severely injured: Genjian, *Haladzagani Husher*, 23.

The stench: Sos-Vani, "The Deportation," 708.

70 *ten days after the Armenians' jailing*: Ishkhanian, "Transcribed from His Journals," 141.

released the captives: Sos-Vani, "The Deportation," 708.

The men were ecstatic: Ishkhanian, "Transcribed from His Journals," 141.

except a dozen: Genjian, *Haladzagani Husher*, 26.

"They did this": Ibid., 27.

"pictured with bombs": Ibid. Genjian gives the date of the bomb photo as Thursday,

July 16, 1915, according to the Julian calendar. Photo is also included in *Badmaqirk*, 722.

hanged in a public square: Aguni, *Milion Me Hayeru Charti Badmutiune*, 269.

In early August: Ibid., 270, gives August 11, 1915, as the deportation's beginning. Hrant Sarian, a teenager from Adabazar who kept a diary, dates the deportation's start as slightly earlier, on Sunday, July 27, 1915 (presumably by the Julian calendar), in "Journal d'un jeune déporté arménien," in *Imprescriptible: Base documentaire sur le génocide arménien,* http://www.imprescriptible.fr/dossiers/temoins/sarian.

"I have very bad news": Vartouhi Boghosian-Gamian, interview by UCLA Armenian Genocide Oral History Project, n.d., trans. Arthur Asatryan.

broke into tears: Kaloustian interviews.

never seen her like this: Ibid.

two long braids: Ibid.

didn't worry her daughter: Ibid.

"Temporary Law of Deportation": Dadrian, *The History of the Armenian Genocide,* 221–22.

"Army, independent corps": Yervant Odian, *Accursed Years: My Exile and Return from Der Zor, 1914–1919,* trans. Ara Stepan Melkonian (London: Gomidas Institute, 2009), 23.

71 *first to leave:* Kaloustian interviews.

establish their businesses: Sos-Vani, "The Deportation," 709.

everything on the sidewalk: Holt, "Fuller Statement," 402.

fifty rugs: Dikran Kamjian, handwritten memoir (1922), 6; manuscript provided by the historian Vahakn Dadrian.

coins in belts: Armaghanian, *Arsha's World and Yours,* 12.

"All was very quiet": Holt, "Fuller Statement," 402–3.

72 *swept the minority Christians:* RK, 263.

Protestant and Catholic Armenians: Turkish Republic Prime Ministry General Directorate of the State Archives, "Armenians Not Subject to Deportation," in *Armenians in Ottoman Documents,* 11.

Greeks were deported: Henry Morgenthau, *United States Diplomacy on the Bosphorus,* ed. Ara Sarafian (Reading, England: Taderon, 2004), 54.

converted to Islam: Turkish Republic Prime Ministry General Directorate of the State Archives, "Deportation of Armenians Who Converted to Islam," in *Armenians in Ottoman Documents,* 63–64.

eyewitness reports: Official Records, 126, 130–31, 137.

two years into his post: Morgenthau, *United States Diplomacy on the Bosphorus,* 4.

"Persecution of Armenians": Official Records, 51–52.

irrelevant to American issues: Samantha Power, *A Problem from Hell: America and the Age of Genocide* (New York: Basic Books, 2002), 8.

"Have you received my 841": Official Records, 55.

joined ninety other Armenian laborers: SM3, 16–18.

73 *Armenian seminary:* Zaven Der Yeghiayan, *My Patriarchal Memoirs* (Barrington, RI: Mayreni, 2002), 87, 90.

74 *"Make absolutely certain":* SM3, 15–16.
kissing the house: Kamjian memoir, 6.
left his door unlocked: Mavian, "The Deportation," 711.
"God listens to children's prayers": Mannig Kouyoumjian, unpublished journals, provided to the author by her daughter Aida Kouyoumjian in May 2008.
"They may have bombs": Holt, "Fuller Statement," 403.
three stone steps: Kaloustian interviews.

75 *trucks barreled:* Holt, "Fuller Statement," 403.
the town of Arifiye: Kaloustian interviews.
bohchas: Ibid.
rides were not free: Rev. A. Hallner, *Uncle Sam: The Teacher and the Administrator of the World* (Sacramento: News Publishing Company, 1918), 120.
paid for the Miskjians: Kaloustian interviews.
their long walk: Holt, "Fuller Statement," 403.
bottlenecked at the station: Ibid., 404.
into the unknown: Kaloustian interviews.

Under the Black Tree

76 *beautiful water mill:* Osman Köker, ed., *Armenians in Turkey 100 Years Ago: With the Postcards from the Collection of Orlando Carlo Calumeno* (Istanbul: Birzamanlar Yayıncılık, 2005), 113.
charted the six mosques: "Regional Statistics: Adabazar," *Piût'ania,* 32 (December 29, 1911): 538–9.

77 *predominantly Muslim:* Hemşehri [Fellow Countrymen] (Sakarya Bükükşehir Belediyesi, February 2011), 15.

78 *Adapazarı had its own museum:* "Adapazarı Depremeleri," Adapazarı Deprem ve Kültür Müzesi, August 2007.

81 *flood the town:* "Many Drown in Floods," *Washington Post,* April 15, 1907; *Badmaqirk,* 420.

Night Train

83 *prodded them outside:* SM3, 19–36.
never known thirst like this before: Anahid MacKeen's memories from conversations with her father.

84 *swollen and split:* Henry Morgenthau, *Ambassador Morgenthau's Story* (Garden City, NY: Doubleday, Page, 1919), 306.

85 *"There were no seats":* Kaloustian interviews.
"To assure their comfort": "Official Proclamation," June 28, 1915, in *Official Records,* 134.

"Where are we going?": Morgenthau, *Ambassador Morgenthau's Story*, 310.

captures an image: Archival signature HADB, Or 1704, Deutsche Bank AG, Historisches Institut.

forced out of their homes: George Horton, November 8, 1915, in *Official Records*, 378–79.

unfamiliar names: RK, 263.

even the sickly: "Letter, Dated 3/16th August, 1915, Conveyed Beyond the Ottoman Frontier by an Armenian Refugee from Cilicia in the Sole of Her Shoe," in *The Treatment of Armenians in the Ottoman Empire*, ed. James Viscount Bryce and Arthur J. Toynbee (1916; reprint, Beirut: G. Doniguian and Sons, 1989), 20.

death march: Morgenthau, *Ambassador Morgenthau's Story*, 311–12.

"Even veiled women": Mihran Mavian, "The Deportation," in *Badmaqirk*, 712.

86 *"Turkish anti-Armenian activities"*: Henry Morgenthau, August 11, 1915, in *Official Records*, 77–78.

93 percent: Ara Sarafian, *Talaat Pasha's Report on the Armenian Genocide, 1917* (London: Gomidas Institute, 2011), 26.

figures of the interior minister: Ibid.

"last days of their lives": Zaven Der Yeghiayan, *My Patriarchal Memoirs* (Barrington, RI: Mayreni, 2002), 87.

leaders taken on April 24: Mikayel Shamtanchian, *The Fatal Night: An Eyewitness Account of the Extermination of Armenian Intellectuals in 1915*, trans. Ishkhan Jinbashian (Studio City, CA: H. and K. Manjikian, 2007), 5–6.

bodies dumped overboard: Morgenthau, *Ambassador Morgenthau's Story*, 312.

muhajirs: Ibid., 311.

wealth had been requisitioned: "Aleppo: Series of Reports from a Foreign Resident at Aleppo, Communicated by the American Committee for Armenian and Syrian Relief," in *The Treatment of Armenians in the Ottoman Empire*, 549.

lines of five thousand: Tashnagtsutiun Committee, "The Extermination of the Armenian People," in *Official Records*, 156, 161.

87 *the rails were overloaded*: Bryce and Toynbee, *The Treatment of Armenians in the Ottoman Empire*, 484–85.

without toilets: Kaloustian interviews.

station towered several stories: Köker, *Armenians in Turkey*, 110.

masses of Armenians: Armenuhi Manuelian, *Dzaghigner Dzaghgamanneres* [Flowers from My Flower Pots] (Boston: Shirag Publishing, 1960), 347–48.

crude tents: Ibid.

open fires: Yervant Odian, *Accursed Years: My Exile and Return from Der Zor, 1914–1919*, trans. Ara Stepan Melkonian (London: Gomidas Institute, 2009), 55.

air of a festival: Ibid.

transient population had swelled: Wilfred M. Post, September 3, 1915, in *Official Records*, 246.

thirty to forty: Ibid.

explained to the American ambassador that August: Henry Morgenthau, *United States Diplomacy on the Bosphorus,* ed. Ara Sarafian (Reading, England: Taderon, 2004), 297–98.

"In the first place": Morgenthau, *Ambassador Morgenthau's Story,* 337.

"It is no use": Ibid., 337–38.

the family was pushed onward: Kaloustian interviews.

88 *Fifteen hundred houses: Annuaire Oriental, 1914: Commerce, Industrie Administration, Magistrature de L'Orient* (Constantinople: Autrichienne Ferd. Walla, 1914), 1662.

their new home: Kaloustian interviews.

Pleas to save the Armenians: William S. Dodd, "American Hospital at Konia, Turkey," in *Official Records,* 192–95; J. B. Jackson, August 10, 1915, in ibid., 198–200.

German ambassador did raise the subject: Henry Morgenthau, August 20, 1915, in ibid., 124.

interfere in their domestic affairs: Samantha Power, *A Problem from Hell: America and the Age of Genocide* (New York: Basic Books, 2002), 13–14.

"Turks Are Evicting": "Turks Are Evicting Native Christians," *New York Times,* July 12, 1915.

"Wholesale Massacres": "Wholesale Massacres of Armenians by Turks," *New York Times,* July 29, 1915.

"Report Turks Shot": "Report Turks Shot Women and Children: Nine Thousand Armenians Massacred and Thrown into Tigris, Socialist Committee Hears," *New York Times,* August 4, 1915.

American Committee for Syrian and Armenian Relief: "Who We Are: History," "Timeline: Highlights," Near East Foundation, accessed on November 20, 2014, http://www.neareast.org

89 *Great Crime:* Vartan Matiossian, "The Birth of 'Great Calamity': How 'Medz Yeghern' Was Introduced onto the World Stage," *Armenian Weekly,* October 25, 2012.

90 *the forward momentum:* SM4, M/1–M/2.

91 *"Destruction of Armenian race":* Henry Morgenthau, September 3, 1915, in *Official Records,* 147.

but wouldn't succumb: SM3, 31–36.

"We made a hole in the ground": Kaloustian interviews.

"no notion of time": Ibid.

92 *a head of hair:* Ibid.

only a few girls had been snatched: Wilfred Post, in *Official Records,* 247.

divvied up by looks: "Armenian Women Put Up at Auction," *New York Times,* September 29, 1915.

Stepan's sister Aghavni came down: Kaloustian interviews.

"Aghavni couldn't give milk": Ibid.

93 *home to fifty thousand residents: Annuaire Oriental 1914: Commerce, Industrie Administration, Magistrature de L'Orient* (Constantinople: Annuaire Oriental, 1914), 1534.

94 *two weeks' wages:* Suleyman Ozmucur and Şevket Pamuk, "Real Wages and the

Standards of Living in the Ottoman Empire, 1489–1914," *The Journal of Economic History*, 62, no.2 (June 2002): 293–321.

eastward on the Baghdad Railway: Edward Mead Earle, *Turkey, The Great Powers and the Bagdad Railway: A Study in Imperialism* (New York: Macmillan, 1924), 71–72.

burrow a tunnel: Ibid., 288–89.

Across a plain: H. Charles Woods, "The Baghdad Railway and Its Tributaries, *The Geographical Journal* 50, no. 1 (July 1917), 40.

holding its breath through the mountains: Bryce and Toynbee, *The Treatment of Armenians in the Ottoman Empire*, 451.

Bottomless gorges: Walter M. Geddes, "Memorandum," November 8, 1915, in *Official Records*, 383.

The cost was twenty ghurush: SM4, 1a.

five hundred thousand Armenians: Wilfred Post, November 25, 1915, in *Official Records*, 396.

95 *"The valley was strewn with graves"*: Ibid.

10 percent were reaching Deir Zor: Ibid.

Lampron: William Francis Ainsworth, *A Personal Narrative of the Euphrates Expedition*, vol. 1 (London: Kegan Paul, Trench, 1888), 150.

ten thousand Armenians wasted away: Dr. L., "The Taurus and Amanus Passes," in *The Treatment of Armenians in the Ottoman Empire*, 454.

"the sad view of the Armenian deportees": Colmar Freiherr von der Goltz, in *The Ottoman Army: Disease & Death on the Battlefield, 1914–1918*, Hikmet Özdemir (Salt Lake City: University of Utah Press, 2008), 138.

96 *dropped to the ground*: Anahid Miskjian, memory of conversations with Stepan Miskjian.

Lusintak, or Moonbeam: Armine Liberatore, granddaughter of Tevon Harutunian, interview with author, Santa Clarita, California, June 29, 2007.

sick with scarlet fever: Carol Neston, Louise's daughter, interview with the author, November 19, 2014.

Where is Stepan: YDZ1, 76–77.

97 *across his throat*: Kaloustian interviews.

"Talaat [Pasha] was going to come": Ibid.

worried their diseases would spread: Turkish Republic Prime Ministry General Directorate of the State Archives, "Armenians to be Deported from Karahisar," in *Armenians in Ottoman Documents, 1915–1920* (Ankara: Başbakanlık Devlet Arşivleri Genel Müdürlüğü, 1995), 108–9.

two thousand Armenians at Chai: Turkish Republic Prime Ministry General Directorate of the State Archives, "Deportation of Armenians from Karahisar and Çay," in *Armenians in Ottoman Documents, 1915–1920* (Ankara: Başbakanlık Devlet Arşivieri Genel Müdürlüğü, 1995), 103–4.

bribed the guards: Kaloustian interviews.

split up once more: Ibid.

The Interior

99 *full of mosques, and no gardens:* Kaloustian interviews.
100 *make membership contingent:* Elizabeth Kolbert, "Dead Reckoning," *New Yorker*, November 6, 2006.
    poor record on civil and political rights: "Commission Staff Working Document: Turkey 2006 Progress Report," Brussels, August 11, 2006, 13–15, http://ec.europa.eu /enlargement/pdf/key_documents/2006/nov/tr_sec_1390_en.pdf.
101 *discrimination against this minority:* Mesut Yeğen, "'Jewish-Kurds' or the New Frontiers of Turkishness," *Patterns of Prejudice* 41 (2007): 1–20.
    role in the persecution of others: RK, 324, 399.
    ever be Turkish enough: Yeğen, "'Jewish-Kurds.'"
    "Opium Black Castle": Josiah Conder, *The Modern Traveller: Syria and Asia Minor,* vol. 2 (London: J. Moyers, 1918), 314.
102 *"Man should always be ready":* SM1, 8.

Infidel Mountains

103 Our situation is deteriorating: SM3, 35–44.
    warmer climate than the mountains: "Q.: Letter," November 25, 1915, in *The Treatment of Armenians in the Ottoman Empire,* ed. James Viscount Bryce and Arthur J. Toynbee (1916; reprint, Beirut: G. Doniguian and Sons, 1989), 436.
    cotton fields: Mihran Mavian, "The Deportation," in *Badmaqirk,* 715.
104 *twenty thousand of them languished:* Anonymous, "Letter from Dr. L.," in *The Treatment of Armenians in the Ottoman Empire,* 454.
    "The stream of deported Armenians": "AE., a Town on the Railway: Series of Reports from a Foreign Resident at AE., Communicated by the American Committee for Armenian and Syrian Relief," in ibid., 453.
105 *no idea where they were headed:* SM4, 1a–2a.
    half a mile in both directions: Walter M. Geddes, "Memorandum," November 8, 1915, in *Official Records,* 380.
106 *one-third had enough food:* "AE., a Town on the Railway," 452.
    countless numbers suffered from typhus: Dr. L., "The Taurus and Amanus Passes," in *The Treatment of Armenians in the Ottoman Empire,* 454.
    "Thousands of exiles": Paula Schäfer, January 18, 1916, in *Official Records,* 435.
    "no foreigners should help": Henry Morgenthau, *Ambassador Morgenthau's Story* (Garden City, NY: Doubleday, Page, 1919), 359–62.
    Amanus Mountains: Grigoris Balakian, *Armenian Golgotha,* trans. Peter Balakian and Aris Sevag (New York: Alfred A. Knopf, 2009), 220.
107 *for the Baghdad Railway workers:* Ibid., 222–23.
    "Hanum": Sister B. Rohner, "Report of Sister B. Rohner About a Visit in the Camp of Mamouret on November 26, 1915," in *Official Records,* 435.

trains needed the wood: Balakian, *Armenian Golgotha,* 241.

Completing the tunnel was difficult: H. Charles Woods, "The Baghdad Railway and Its Tributaries," *The Geographical Journal* 50, no. 1 (July 1917), 46.

lacked a natural a passage: Edward Mead Earle, *Turkey, the Great Powers, and the Bagdad Railway: A Study in Imperialism* (New York: Macmillan, 1924), 72.

costliest efforts: Ibid., 22.

An opening in September: Ibid., 289.

Kanlï-Gechid: Balakian, *Armenian Golgotha,* 222–23.

remains of hundreds of thousands: Ibid., 223.

108   *Infidel Mountains*: Ibid., 228–29.

stood an Armenian church: Ibid., 230.

seven hours by foot: Ibid., 230–31.

The Headscarf

112   *visited outside Adapazari*: Osman Köker, ed., *Armenians in Turkey 100 Years Ago: With the Postcards from the Collection of Orlando Carlo Calumeno* (Istanbul: Birzamanlar Yayıncılık, 2005), 117.

113   *operations to the Middle East*: Sebnem Arsu and Brian Knowlton, "Planned House Vote on Armenian Massacre Angers Turks," *New York Times,* March 30, 2007.

cut off access to the base: Desmond Butler, "Congressional Panel OKs Armenian Measure," *Washington Post,* October 10, 2007.

momentum for official recognition: Aram S. Hamparian, "Support for Armenian Genocide Resolution Growing Among Key National Security Committees," press release, Armenian National Committee of America, April 10, 2007.

promised to move a vote: Carl Hulse, "U.S. and Turkey Thwart Armenian Genocide Bill," *New York Times,* October 26, 2007.

George W. Bush was vigorously opposing it: Ryan Grim, "Pols Sidestep Debate Over Armenian Genocide," *Politico,* April 24, 2007.

eight former secretaries of state: Glenn Kessler, "White House and Turkey Fight Bill on Armenia," *Washington Post,* October 10, 2007.

"Unfortunately, some politicians": Butler, "Congressional Panel OKs Armenian Measure."

the resolution died: Hulse, "U.S. and Turkey Thwart Armenian Genocide Bill."

land once owned by Armenians: Carol Williams, "Descendants of Armenian Genocide Victims Seek $65 Million from Turkey for Seized Land," *Los Angeles Times,* December 15, 2010.

114   Blessings and abundance: Translation and analysis done by Alidz and Hrant Agbabian.

guiding Armenian tourists: Jemal (Cemal) worked at the California-based Armenian Heritage Society Tours; phone interview with founder Armen Aroyan, December 8, 2014.

Dreams Traded for Bread

115 *a mass of people:* SM3, 44–74.

tens of thousands: Hoffman, "Enclosure 1," 1916-01-03-DE-001, "Armenian Genocide."

a rainbow: Mannig Kouyoumjian, unpublished journals, provided to the author by her daughter Aida Kouyoumjian in May 2008.

blankets: Vahram Dadrian, *To the Desert: Pages from My Diary,* trans. Agop J. Hacikyan (Reading, England: Taderon, 2003), 51.

to save the Armenians: J. B. Jackson, September 29, 1915, in *Official Records,* 307–8; Dr. Martin Niepage, *The Horrors of Aleppo . . . Seen by a German Eyewitness* (London: T. Fisher Unwin, 1916), 5–9.

discarded on roadsides: Dadrian, *To the Desert,* 56.

116 *military depot:* Djemal Pasha, *Memories of a Turkish Statesman, 1913–1919* (New York: George H. Doran, 1922), 143.

small glows of fires: Kouyoumjian journals

maze of tents: Dadrian, *To the Desert,* 54.

only patches of thistle: Hoffman, "Enclosure 1."

brother-in-law to the late Tevon: SM3, 7.

strewn upon the ground: Anonymous enclosure, 1915-11-08-001, "Armenian Genocide."

raw sewage: Ibid.

117 *No latrine:* Hoffman, "Enclosure 1."

the stench: Dadrian, *To the Desert,* 51–52.

grouping themselves that way: Sos-Vani, "Forced Exile," in *Badmaqirk,* 720–23.

rag carpet: YDZ2, 72.

clothes: SM3, 51.

forty-five-minute hike: SM4, 3a.

118 *empire had become overextended:* Liman von Sanders, *Five Years in Turkey* (Baltimore: Williams and Wilkins, 1928), 326.

crossroads of the Baghdad Railway: Djemal Pasha, *Memories of a Turkish Statesman,* 143.

119 *golden stalks:* SM4, 3a.

the population swelling: Walter M. Geddes, "Memorandum," November 8, 1915, in *Official Records,* 381.

"Kalkïn! Chïkïn! Yïkïn": Mihran Mavian, "The Deportation," in *Badmaqirk,* 716.

herded southward toward Aleppo: Enclosure, 1915-11-01-DE-001, "Armenian Genocide."

banks now dotted with Armenian encampments: Anonymous enclosure, 1915-11-16-DE-002, "Armenian Genocide."

120 *sit on a hot copper pot:* Badmaqirk, 515.

animal carcasses: Anonymous enclosure, 1915-11-08-001, "Armenian Genocide."

typhus: Elmasd Santoorian, "A Nurse's Odyssey: From Marash to Aleppo and Back," in *The Cilician Armenian Ordeal,* ed. Paren Kazanjian (Boston: Hye Intentions, 1989), 447.

dysentery: Elise Hagobian Taft, *Rebirth: The Story of an Armenian Girl Who Survived the Genocide and Found Rebirth in America* (Plandome, NY: New Age Publishers, 1981), 57.

fifty to sixty new bodies: SM4, 3a.

rose to two hundred: SM3, 50.

heaped them in oxcarts: Mavian, "The Deportation," 716.

"may have looked like Hell": Santoorian, "A Nurse's Odyssey," 445.

123 *Gallipoli campaign was finally ending:* Robin Prior, "The Ottoman Front," in CHFWW, 311–12.

secretive retreat: Alan Moorehead, *Gallipoli* (New York: HarperCollins, 2002), 334–40.

half a million casualties: John H. Morrow Jr., "The Imperial Framework," in CHFWW, 419–20.

one hundred thousand men who died: Edward J. Erickson, *Ordered to Die: A History of the Ottoman Army in the First World War* (Westport, CT: Greenwood, 2001), 94.

Armenians dreamed would rescue them: Alexander Aintablian, "At Last 'Uncle' Came," in *The Cilician Armenian Ordeal,* ed. Paren Kazanjian (Boston: Hye Intentions, 1989), 48.

"I have accomplished more": Henry Morgenthau, *Ambassador Morgenthau's Story* (Garden City, NY: Doubleday, Page, 1919), 342.

124 *bodies clogged a section of the Euphrates River:* Ibid., 318.

the essential words: Dadrian, *To the Desert,* 55.

126 *the five Armenians set out:* SM3, 70–73; SM4, 5a–5b.

comparable to Adabazar: Annuaire Oriental 1914: Commerce, Industrie Administration, Magistrature de L'Orient (Constantinople: Annuaire Oriental, 1914), 1528.

houses rose several stories: Author's 2007 trip to Kilis; Osman Köker, ed., *Armenians in Turkey 100 Years Ago: With the Postcards from the Collection of Orlando Carlo Calumeno* (Istanbul: Birzamanlar Yayıncılık, 2005), 286–87.

He didn't see Armenians: SM4, 5b.

twenty-six hundred had resided: Annuaire Oriental, 1528.

several of the old structures: Ibid.; "Hamamlar," T. C. Kilis Valiliği, http://www.kilis. gov.tr.

128 *vesika:* J. B. Jackson, "Subject: Armenian Atrocities," in *Official Records,* 594.

protect passing troops: Rössler, 1916-02-09-DE-001, "Armenian Genocide."

129 *Muslimiye:* SM4, 7a.

The Bath

131 *three and a half centuries:* "Hamamlar," T. C. Kilis Valiliği, http://www.kilis.gov.tr.

132 *last Armenian village, Vakıflı Köyü:* Matthew Brunwasser, "In Turkey's Last Armenian Village, a Place to Get Away from It All," *PRI's The World,* December 28, 2011.

Water's Course

134 *his caravan walked southward:* SM4, 7a–17b.

   "An ample supply of provisions": Karl Baedeker, *Palestine and Syria, with Routes Through Mesopotamia and Babylonia and the Island of Cyprus* (Leipzig: Karl Baedeker, 1912), 435.

   a band of robbers: Hovhannes Khatcherian, "A Description of Bab," in *L'Extermination des Déportés Arméniens Ottomans dans les Camps de Concentration de Syrie-Mésopotamie (1915–1916), la Deuxième Phase du Génocide* [The Extermination of Ottoman Armenians Deported to the Concentration Camps of Syria-Mesopotamia], ed. Raymond H. Kévorkian (Bibliotheque Nubar de l'Union Générale Arménienne de Bienfaisance, 1998), 75.

135 *clay field:* RK, 635.

   once every decade: Aram Andonian, "Bab," in *L'Extermination,* 81.

   typhus: RK, 635.

136 *inundated with accounts:* Henry Morgenthau, *Ambassador Morgenthau's Story* (Garden City, NY: Doubleday, Page, 1919), 381.

   leave his post: Ibid., 390.

   his farewell meeting: Ibid., 390–92.

   stalemated: Robin Prior, "1916: Impasse," in *CHFWW,* 89.

   a devastated Serbia: Morgenthau, *Ambassador Morgenthau's Story,* 395–96.

   to Berlin: Ibid., 393.

   Wilson prepared to run for reelection: Ibid., 385.

   American public read about the war: "Mesopotamia Battle Amid a Hurricane," *New York Times,* January 22, 1916.

   offensive on the Caucasus front: Franz Carl Endres, "Turkish Campaigns: Operations on the Caucasus Front," in *Encyclopaedia Britannica,* 12th ed., vol. 12 (London: Encyclopaedia Britannica Company, 1922), 804–5.

   "My failure to stop": Morgenthau, *Ambassador Morgenthau's Story,* 385.

137 *five hundred perished:* Andonian, "Bab," 80.

   thousand others died within just sixty hours: Rössler, 1916-02-09-DE-001, "Armenian Genocide."

   corpses by the feet: Andonian, "Bab," 88.

   Not wanting images of this circulating: Guerini E Associati, *Armin T. Wegner and the Armenians in Anatolia, 1915: Images and Testimonies* (Milan: Guerini e Associati, 2007), 35.

138 *smuggled out his photographs:* Ibid., 35–37.

   paper-thin clothes: Ibid., 117, 125.

   five hundred thousand Armenians wasted away: "Aleppo: Series of Reports From a Foreign Resident at Aleppo; Communicated by the American Committee for Armenian and Syrian Relief," in *The Treatment of Armenians in the Ottoman Empire,* ed. James Viscount Bryce and Arthur J. Toynbee (1916; reprint, Beirut: G. Doniguian and Sons, 1989), 550.

warning to the German ambassador: Robert Lansing, "The Secretary of State to the German Ambassador (Bernstorff)," February 16, 1916, in *Official Records*, 453.

Was that an aircraft: Andonian, "Bab," 80–81.

it would continue until 1923: "Impressive Local Demonstration," *New York Times*, July 25, 1923.

yet another move: SM4, 10a–10b; Rössler, 1916-02-09-DE-001, "Armenian Genocide."

139 *close the encampments near Aleppo:* RK, 636.

"Second Phase": Ibid., 693.

Greek Orthodox Christians, ousted: J. B. Jackson, "Armenian Atrocities," in *Official Records*, 595.

previously excluded Armenian: RK, 664.

10 percent of the local Muslim population: Rössler, 1916-04-27-DE-001, "Armenian Genocide."

Ali Suad Bey: A. Bernau, September 10, 1916, in *Official Records*, 559.

captured Erzerum: Liman von Sanders, *Five Years in Turkey* (Baltimore: Williams and Wilkins, 1928), 124.

once flourishing with Armenians: "15,000 Massacred as Erzerum Fell," *New York Times*, May 3, 1916.

Russians exercised rough justice: "Russians Slaughter Turkish Third Army," *New York Times*, March 6, 1916.

There is a higher law: YDZ1, 20.

140 *wild dogs:* Khatcherian, "A Description of Bab," 77.

holding their children's hands: Donald E. Miller and Lorna Touryan Miller, *Survivors: An Oral History of the Armenian Genocide* (Berkeley: University of California Press, 1999), 105.

around its banks: Baedeker, *Palestine and Syria*, 415–17.

The riverside town of Meskene: SM4, 11a.

an estimated ten thousand denizens: Wilhelm Litten, February 6, 1916, enclosure, 1916-02-09-DE-001, "Armenian Genocide."

swell to sixty thousand: Bernau, in *Official Records*, 557.

a path that shadowed the Euphrates: SM4, 11a–17b.

141 *small islands:* Baedeker, *Palestine and Syria*, 435.

Ottoman army had surrounded: Stéphane Audoin-Rouzeau, "1915: Stalemate," in CHFWW, 73.

"bleed France white": "Forty Years and Forty Weeks," *New York Times*, February 20, 1917.

stalemates: Prior, "1916: Impasse," in CHFWW, 108–9; and Robin Prior, "The Western Front," in CHFWW, 213–15.

approaching two million men: Prior, "1916: Impasse," 108.

Sussex: John Bach McMaster, *The United States in the World War* (New York and London: D. Appleton, 1918), 215–16, 229.

sever diplomatic ties: Ibid., 323.

142 *two-thousand-plus:* RK, 657.

Alexander the Great: Baedeker, *Palestine and Syria,* 435.
peering at the gravedigger's ledger: SM4, 16a.
escalated the sevkiyat: RK, 657.
"I have seen with my own eyes": Wilhelm Litten, February 6, 1916, enclosure, 1916-02-09-DE-001, "Armenian Genocide."

143 *blades of grass:* SM4, 11a.
144 *The convoy:* YDZ1, 23–35.
      barley seed: SM4, 14a; Preacher Vartan Geranian, "Enclosure 4," June 28, 1916, 1916-07-29-DE-001, "Armenian Genocide."
146 *left bank of the Euphrates:* RK, 658.
      a severe penalty: Krikor Ankout, "Devant Rakka," in *L'Extermination,* 158.
      Raqqa was swollen with refugees: RK, 661.
      safe haven: Bernau, in *Official Records,* 558.
147 *houses of crimson brick:* Ankout, "Devant Rakka," 159–60.
      money through the mail: Ibid., 165.
      disappeared like ghosts: Baedeker, *Palestine and Syria,* xlix.
148 *"return you to your villages":* YDZ1, 38–39.
      Arab Revolt: Djemal Pasha, *Memories of a Turkish Statesman, 1913–1919* (New York: George H. Doran, 1922), 168.
      declaring his independence: David Fromkin, *A Peace to End All Peace* (New York: Henry Holt, 1989), 218.
149 *bouquet of promises:* Robin Prior, "The Ottoman Front," in CHFWW, 316–17.
      Ras al-Ayn had become a slaughterhouse: J. B. Jackson, "Armenian Atrocities," in *Official Records,* 591; RK, 651–52.
      "Walking and walking, my legs": Verjiné Svazlian, *The Armenian Genocide: Testimonies of the Eyewitness Survivors* (Yerevan: Gitut'yun Publishing, 2011), 570.
      water does not forget its course: R. G. Bayan, trans., *Armenian Proverbs and Sayings* (Venice: Academy of St. Lazarus, 1889), 25.
      names would become imprinted: YDZ1, 34–35.
150 *thirty-two-year-old Harutiun:* Mgrdich Bodurian, ed., *Qaghut'ahay Dareqirk* [Diasporan Armenian Yearbook] (Bucharest: Hay Mamul, 1940), 141.
      "horrific parade of bodies": Wilhelm Litten, February 6, 1916, enclosure, 1916-02-09-DE-001, "Armenian Genocide."

The Dead Zone

153 *Marqada:* H. G. Atanasian, "The Last Act of the Deir Zor Massacre: An Excerpt from My Memoirs," *Hay Mamul,* Bucharest, Romania, no. 38 (267), April 28, 1940. Also, a description of area is given in H. G. Atanasian, "The Exiled Life: Separation," in *Rumanahay Daretsuyts 1925* [Romanian Armenian Yearbook], ed. H. G. Adrushan (Bucharest: H. M. Bodurian, H. G. Atanasian, 1924), 137.
154 *didn't want to take any chances:* Maureen Freely, "I Stand By My Words," *Guardian,* October 22, 2005.

sheltered my grandfather: YDZ2, 31–66.

United States warned travelers: U.S. Department of State, Bureau of Consular Affairs, "Travel Warning: Syria," November 13, 2006.

sponsoring terrorism: U.S. Department of State, Bureau of Diplomatic Security, "Travel Warning: Syria," April 17, 2008.

armed men: Rhonda Roumani, "Four Armed Men Attack U.S. Embassy in Damascus," *Washington Post*, September 13, 2006.

killing one person and injuring thirteen: Craig S. Smith, "Survivor of Attack on U.S. Embassy Dies in Syria," *New York Times*, September 13, 2006.

Hell

157 *"birds flew away from the trees"*: Verjiné Svazlian, *The Armenian Genocide: Testimonies of the Eyewitness Survivors* (Yerevan: Gitut'yun Publishing, 2011), 568.
Deir Zor: YDZ1, 36–45; SM4, 17b–19a.
spilling outward from the riverbank: Mannig Kouyoumjian, unpublished journals, provided to the author by her daughter Aida Kouyoumjian in May 2008.
begging for food: "The Journey Report of Our Committee Members," enclosure, 1916-07-22-DE-002, "Armenian Genocide."
Multiple generations: Kouyoumjian journals

158 *"As on the gates of 'Hell'"*: A. Bernau, September 10, 1916, in *Official Records*, 556.
interred in mass graves: Kouyoumjian journals.
abuzz with chatter: SM4, 17b.
get the wheel fixed: YDZ1, 36.
towered a mosque: Yervant Odian, *Accursed Years: My Exile and Return from Der Zor, 1914–1919*, trans. Ara Stepan Melkonian (London: Gomidas Institute, 2009), 188.

159 *new governor threatened to hang*: Sebuh Aguni, *Milion Me Hayeru Charti Badmutiune* [History of the Massacre of a Million Armenians] (Constantinople: H. Asadourian and Sons, 1921), 315.

160 *"don't believe everything"*: YDZ1, 39–41.
Zeki Bey: Aguni, *Milion Me Hayeru Charti Badmutiune*, 315.

161 *totaled a month*: YDZ1, 43–45.
"persecution of the Armenians": Paul Wolff Metternich, 1916-07-10-DE-001, "Armenian Genocide."

162 *legions more*: "Notes by the Consul General," 1915-11-06-DE-012, "Armenian Genocide"; "From the Ambassador in Extraordinary Mission in Constantinople," 1916-04-03-DE-002, "Armenian Genocide"; "From the Consul in Damascus," 1916-05-30-DE-001, "Armenian Genocide"; "From the Consul in Aleppo," 1916-06-29-DE-001, "Armenian Genocide."
Speculation in Aleppo: Martin Niepage, *The Horrors of Aleppo . . . Seen by a German Eyewitness* (London: T. Fisher Unwin, 1916), 8–9.
"Not a leaf would turn": YDZ1, 20.

"*Germany can never consent*": Zaven Der Yeghiayan, "From the Armenian Patriarch in Constantinople," 1916-05-12-DE-001, "Armenian Genocide."

batches of two to five thousand prisoners: RK, 665.

In the middle of July: Araxia Jebejian, July 17, 1916, letter, 1916-07-29-DE-001, "Armenian Genocide."

arrested the remaining priests: Ibid.

"*dangerous to the health*": "From the Consul in Aleppo," 1916-06-29-DE-001, "Armenian Genocide."

On July 22: Rössler, 1916-07-29-DE-001, "Armenian Genocide."

"*The Minister admits*": Hoffman Philip, July 26, 1916, in *Official Records*, 529–30.

163   human flood: YDZ1, 45–72; SM4, 19a–24a.

the end of world: Anahid MacKeen's memories of her father.

164   *black spectacles*: Dikran Jebejian, *Abrvadz Orer Hayasbanutene'n* [Days Lived Through the Armenocide] (Aleppo: Cilician Bookstore Publishing and Printing House, 2001), 8.

deep-set eyes: Ernest C. Partridge, "The Pensacola Party and Relief Work in Turkey," *Armenian Affairs* 1 (Summer and Fall 1950), unnumbered photo.

The twenty-eight-year-old official: Jebejian, *Abrvadz Orer Hayasbanutene'n*, 8.

"*starving Armenians*": "The Repatriation of Armenians," *New Armenia* 8 (July 15, 1916): 250.

"*He kept us out of the war*": Ernest W. Young, *The Wilson Administration and the Great War* (Boston: Gorham Press, 1922), 20.

165   *short in stature*: Jebejian, *Abrvadz Orer Hayasbanutene'n*, 8.

Everek: Aguni, *Milion Me Hayeru Charti Badmutiune*, 315.

166   *changed course*: SM4, 20b.

wounds of water deprivation: Earl of Ronaldshay, *On the Outskirts of Empire in Asia* (Edinburgh: William Blackwood and Sons, 1904), 66.

military station: SM4, 21a.

167   *golden-brown shrivels*: Araxia Jebejian, June 22, 1916, letter, 1916-07-29-DE-001, "Armenian Genocide."

burned his cart in a bonfire: YDZ1, 80.

one thousand children: Ibid., 67–68.

stories came from a boy: Ibid., 53–59.

168   "*Aside from less than a hundred*": J. B. Jackson, "Subject: Armenian Atrocities," March 4, 1918, in *Official Records*, 591.

a naked man: YDZ1, 67–72; SM4, 23b–24a.

Welcome to Syria

172   *thriving community of a hundred thousand*: Roza N. Hovhanesyan, department head, Department of Armenian Communities of North and South America and Australia, Armenia's Ministry of Diaspora, e-mail interview with author, December 18 and 21, 2014.

The Desert's End

174 *"You are being sent back home":* YDZ1, 59–62.

175 *trembling convoy waded out:* Ibid., 63–64.
"Let's die and be saved": Ibid., 61.
summoning God: SM4, 26a–29a.
the most affluent deportees: YDZ1, 72–74.

176 *"The man wants two thousand":* Ibid., 74–75.
"Ahead toward the desert": Ibid., 74–93.

177 *Harutiun Atanasian, saw them:* H. G. Atanasian, "The Last Act of the Deir Zor Massacre: An Excerpt from My Memoirs," *Hay Mamul,* Bucharest, Romania, no. 38 (267), April 28, 1940.
massacre site: SM4, 26a.
great flare illuminated the desert: SM4, 26b.
only Thursday: Atanasian, "The Last Act."

178 *about one hundred feet:* H. G. Atanasian, "The Exiled Life: Separation," in *Rumanahay Daretsuyts 1925* [Romanian Armenian Yearbook], ed. H. G. Adrushan (Bucharest: H. M. Bodurian, H. G. Atanasian, 1924), 138.
village of Shedadiye: YDZ1, 79.
swooshing was loud: Atanasian, "The Exiled Life," 138.
tents of three hundred guards: Ibid., 137.
by using a prop: "Deir-Zor, Zéki," *L'Extermination des Déportés Arméniens Ottomans dans les Camps de Concentration de Syrie-Mésopotamie (1915–1916), la Deuxième Phase du Génocide* [The Extermination of Ottoman Armenians Deported to the Concentration Camps of Syria-Mesopotamia], ed. Raymond H. Kévorkian (Bibliothèque Nubar de l'Union Générale Arménienne de Bienfaisance, 1998), 190.
two hundred thousand Armenians: RK, 662.
six thousand remained: YDZ1, 82.
Two more days of wasting away: Atanasian, "The Last Act." September 24 in the Julian calendar translates to October 7, a Saturday.
Dozens of children: SM4, 27a.
Harutiun thought Deir Zor: Atanasian, "The Last Act."

179 *the drowning of two thousand children:* RK, 667–68.
"We noticed smoke": Aram Bouloutian, interview by UCLA Armenian Genocide Oral History Project, n.d., translated by Ani Aramyan.

180 *October 8:* Atanasian, "The Exiled Life," 137. Miskjian pinpoints his escape as Sunday, September 24, in the Julian calendar, but the twenty-fourth was a Saturday. The manuscript has been corrected to reflect Harutiun's correct account; SM4, 29a.

181 *full moon:* Atanasian, "The Exiled Life," 138.
several hundred families left: Ibid., 137.

182 *pointless to hide behind:* Ibid., 138.

183 *perhaps eight o'clock:* SM4, 28b.

kill all males over age twelve: Atanasian, "The Last Act."

The two began to sob: Atanasian, "The Exiled Life," 140.

three and a half hours: SM4, 28b.

My Shadow

185 *terminus of the deportation:* RK, 662–65.

186 *on the deportation road:* Aram Andonian, "Notes relatives à Deir-Zor," in *L'Extermination des Déportés Arméniens Ottomans dans les Camps de Concentration de Syrie-Mésopotamie (1915–1916), la Deuxième Phase du Génocide* [The Extermination of Ottoman Armenians Deported to the Concentration Camps of Syria-Mesopotamia], ed. Raymond H. Kévorkian (Bibliothèque Nubar de l'Union Générale Arménienne de Bienfaisance, 1998), 184.

hill of bones: "Turkey and Armenia: Bones to Pick," *Economist*, October 8, 2009.

thousands that had died: Yervant Odian, *Accursed Years: My Exile and Return from Der Zor, 1914–1919*, trans. Ara Stepan Melkonian (London: Gomidas Institute, 2009), 156.

Tell the World

188 *across the desert floor:* YDZ1, 93–102; SM4, 29a–31a.

189 *Atanasian wasn't far away:* H. G. Atanasian, "The Exiled Life: Separation," in *Rumanahay Daretsuyts 1925* [Romanian Armenian Yearbook], ed. H. G. Adrushan (Bucharest: H. M. Bodurian, H. G. Atanasian, 1924), 141.

190 *lost their faith:* Donald E. Miller and Lorna Touryan Miller, *Survivors: An Oral History of the Armenian Genocide* (Berkeley: University of California Press, 1999), 179.

why a benevolent God: Jack Tartanian, Arshaluys's son, interview with the author, summer 2006.

Stepan's faith: Anahid MacKeen's recollection of conversations with her father.

masquerading as girls: H. G. Atanasian, "The Last Act of the Deir Zor Massacre: An Excerpt from My Memoirs," *Hay Mamul*, Bucharest, Romania, no. 38 (267), April 28, 1940.

considered worse than death: Anahid MacKeen's recollection of her father.

"I rotted and remained": Verjiné Svazlian, *The Armenian Genocide: Testimonies of the Eyewitness Survivors* (Yerevan: Gitut'yun Publishing, 2011), 573.

193 *four cups of water a day:* Dr. Peter Aronson, Section of Nephrology, Department of Medicine, Yale School of Medicine, e-mail interview with author, Jan. 4, 2015.

194 *"Mirage became the ruling factor":* C. G. Lewis, "A Survey of the Euphrates Valley from Deir-Ez-Zor to Aleppo," *Geographical Journal* 59 (June 1922): 455.

encountering a well: Dikran Jebejian, *Abrvadz Orer Hayasbanutene'n* [Days Lived Through the Armenocide] (Aleppo: Cilician Bookstore Publishing and Printing House, 2001), 22.

"Why should we take the pain?": Ibid.

195 *friends drank from the same fountain:* Ibid.

The Sandstorm

201 Marqada: H. G. Atanasian, "The Last Act of the Deir Zor Massacre: An Excerpt from My Memoirs," *Hay Mamul*, Bucharest, Romania, no. 38 (267), April 28, 1940.

Betrayal

205 *the fourteen naked men:* SM4, 31a–37a; YDZ1, 102–24.

209 *"Armenian and Syrian Relief Days":* "Give Millions Today to Save Armenians," *New York Times*, October 22, 1916.

"in view of the misery": "The Story of the Near East Relief," Armenian Genocide Museum-Institute, National Academy of Sciences of the Republic of Armenia, http://www.genocide-museum.am/eng/photos/06near-larger.jpg.

distribution was still being thwarted: Abram I. Elkus, October 17, 1916, in *Official Records*, 545.

"studied intention": Robert Lansing, November 1, 1916, in ibid., 547.

a wave of killings: "From the Consul in Aleppo," 1916-11-05-DE-001, "Armenian Genocide."

"As far as we know, the annihilation campaign": Ibid.

In a caravan of seventeen hundred: "Enclosure 1," in ibid.

surrounding their caravan: "Enclosure 3," in ibid.

tens of thousands were murdered: Rössler to Bethmann-Hollweg, in ibid.

as high as one and a half million: "From the Chargé d'Affaires in Constantinople," 1916-10-04-DE-002, in ibid.

"interference in the inner relations": "From the Undersecretary of State in the German Foreign Office," 1916-10-08-DE-001, in ibid.

210 *recall its ambassador:* RK, 696.

213 *desert for company:* YDZ2, 3–24.

as big as expected: David Fromkin, *A Peace to End All Peace* (New York: Henry Holt, 1989), 219–25.

capture of some key towns: David Murphy, "Chronology," *The Arab Revolt, 1916–18* (Oxford: Osprey, 2008), 9–10.

a young British officer: Fromkin, *A Peace to End All Peace*, 226.

who would ally: Murphy, *The Arab Revolt*, 36–37.

Feisal ibn al-Hussein: Robin Prior, "The Ottoman Front," in CHFWW, 316.

214 *"Lawrence of Arabia":* Fromkin, *A Peace to End All Peace*, 171.

"Gel, gel": SM4, 37a–39b.

218 *Arab tribesmen:* Murphy, *The Arab Revolt*, 38–39.

More news of Armenian massacres: "The Massacred Armenians," in *The Armenian Genocide: News Accounts from the American Press, 1915–1922*, ed. Richard Diran Kloian (Richmond, CA: Heritage, 2007), 220.

reelected President Woodrow Wilson: "Say Women's Train Bowled Hughes Out," *New York Times,* November 20, 1916.

Battle of Verdun: Robin Prior, "1916: Impasse," in CHFWW, 108.

Battle of the Somme: Ibid.

Beside the Euphrates: SM4, 40a–41a.

The Church

222 *Holy Martyrs Armenian Church:* Sarkis Balmanoukian, phone interview by the author, February 24, 2015.

three walls: Sarkis Balmanoukian, interview by the author, Tarzana, California, July 24, 2007.

friendship wall: Ibid.

Sarkis Balmanoukian: Ibid.

The Sheikh

225 *"Let him go for my sake":* SM4, 41a.

226 *as he fled:* YDZ2, 24–30.

astounded at the turn of events: Anahid MacKeen's remembrance of conversations with her father.

eight hundred skins: William H. Hall, ed., *Reconstruction in Turkey: A Series of Reports Compiled for the American Committee of Armenian and Syrian Relief* (New York: n.p., 1918), 86.

poles to fasten atop: Ibid.

227 *corn stalks:* SM4, M/a–M/b.

ears already harvested: YDZ2, 27.

raft into the water: SM4, M/b.

228 *February morning:* Ibid., M/h.

a throng of black tents: YDZ2, 31–58; "Youssef," Sheikh Hammud al-Aekleh's grandson, interview by the author, Syria, August 31, 2007. Due to the war in Syria, I have changed the names of the sheikh's descendants, including Youssef.

fine hair of goats: "Omar," Sheikh Hammud al-Aekleh's descendant, interview with the author, Syria, August 31, 2007.

"houses of hair": Gertrude Lowthian Bell, *Syria: The Desert and the Sown* (London: William Heinemann, 1928), 23.

a thin round face: "Zeyad," Sheikh Hammud al-Aekleh's grandson, interview with the author, Syria, December 11, 2009.

customary dress for winter: Ibid.

229 *made from grapes:* Hall, *Reconstruction in Turkey,* 151.

230 *two wives:* Youssef interview, August 30, 2007.

barren cliffs down to the Euphrates: "Ahmed," Sheikh Hammud al-Aekleh's descendant, interview with the author, Syria, August 30, 2007.

thirty-five settlements: YDZ2, 66.

232 *forced into harems:* RK, 758.

234 *in the mid-1800s:* Norman N. Lewis, *Nomads and Settlers in Syria and Jordan, 1800–1980* (New York: Cambridge University Press, 2009), 12, 23–29, 35.

238 *slamming a rifle:* SM4, M/h.

Two Hammuds

242 *honored his clan:* "Arab Tribal Leaders Arrive in Yerevan," *Asbarez Armenian News*, March 28, 2005, http://asbarez.com.

Crossroads

244 *livestock thieves:* YDZ2, 61–66.

one of the sheikh's sons: Anahid MacKeen's recollection of conversations with her father.

Baghdad had been lost: "Proclaims Liberation to Citizens of Bagdad," *New York Times*, March 19, 1917.

offensive to reclaim: Liman von Sanders, *Five Years in Turkey* (Baltimore: Williams and Wilkins, 1928), 173–76.

Lightning, or Yildirim: Ibid., 173.

concentrating in the region: Djemal Pasha, *Memories of a Turkish Statesman, 1913–1919* (New York: George H. Doran, 1922), 185.

attack on British positions: Edward Erickson, *Ordered to Die: A History of the Ottoman Army in the First World War* (Westport, CT: Greenwood Press, 2001), 167.

245 *diverting resources:* Djemal Pasha, *Memories of a Turkish Statesman*, 182–85.

gateway to Palestine: Robin Prior, "The Ottoman Front," in CHFWW, 315.

expected yet another British assault: Djemal Pasha, *Memories of a Turkish Statesman*, 183.

Hasan Chavush: YDZ2, 65–80.

revolution sweeping Russia: Michael S. Neiberg, "1917: Global War," in CHFWW, 116.

withdraw from eastern Anatolia: Erickson, *Ordered to Die*, 160–61.

246 *blew about the region:* Homer Davenport, *My Quest of the Arab Horse* (New York: B. W. Dodge, 1909), 237–38.

rose from snowmelt: William H. Hall, ed., *Reconstruction in Turkey: A Series of Reports Compiled for the American Committee of Armenian and Syrian Relief* (New York: n.p., 1918), 85.

other important matters: YDZ2, 50–51.

The sheikh had called on him again: Ibid., 58–61.

247 *the most agonizing decisions:* Anahid MacKeen's memory of her father.

250 *"There's no salvation":* YDZ2, 74.

252 *On the road:* Ibid., 79–99.

253 *Armenian deportees were recruited:* RK, 662.

254 *making it to Raqqa:* H. G. Atanasian, "The Last Act of the Deir Zor Massacre: An Excerpt from my Memoirs," *Hay Mamul,* Bucharest, Romania, no. 38 (267), April 28, 1940.

256 *Americans had joined the Allies:* "President's Proclamation of a State of War and Regulations Governing Alien Enemies," *New York Times,* April 7, 1917.
the Russian Revolution: Neiberg, "1917: Global War," 116–17.
Gaza fell to the British: Prior, "The Ottoman Front," 315–17.
Jerusalem came under British control: "Jerusalem Falls to British Army," *New York Times,* December 11, 1917.

Home

264 *"Help him get on the train!":* YDZ2, 99–100.

265 *The train chugged north and then west:* Ibid., 100–128.
Baghche tunnel: R. J. Bjurstedt, "Germany's Railway Problems in Asiatic Turkey," *Popular Mechanics* (May 1916): 648–50.
the tunnel's hard labor: William H. Hall, ed., *Reconstruction in Turkey: A Series of Reports Compiled for the American Committee of Armenian and Syrian Relief* (New York: n.p., 1918), 99.

266 *deserting in droves:* Ward Price, "Deserter Bands Loot in Austria: Peasants Terrorized and Plundered After the Manner of Mediaeval Days," *New York Times,* August 8, 1918.
more soldiers had fled than remained: Liman von Sanders, *Five Years in Turkey* (Baltimore: Williams and Wilkins, 1928), 243.
contemplate withdrawing their troops: Ibid., 240.
weakened the Central Powers: Christoph Mick, "1918: Endgame," in CHFWW, 145–50, 167.
exhausted troops: von Sanders, *Five Years in Turkey,* 268–69.

267 *directed scarce manpower:* Ibid., 256–57.
reunite the Turkish-speaking people: David Fromkin, *A Peace to End All Peace* (New York: Henry Holt, 1989), 352–54.
over seven hundred thousand: Mustafa Aksakal, "The Ottoman Empire," in CHFWW, 468.
sprawling operation: Photograph provided to author by Ara Chalvardjian, son of Mardiros Chalvarjian, June 1, 2011.

268 *exempted from the deportations:* Vania Chalvardjian, granddaughter of Mardiros Chalvarjian, interview with the author, June 6, 2011.
Chalvarjian brothers: Yervant Odian, *Accursed Years: My Exile and Return from Der Zor, 1914–1919,* trans. Ara Stepan Melkonian (London: Gomidas Institute, 2009), 84.

Mardiros and Khacher Chalvarjian: Ara Chalvardjian, son of Mardiros Chalvarjian, interview with the author, June 1, 2011. (There are different spellings of the family's last name.)

most famous intellectuals: Odian, *Accursed Years*, 84.

269 *Megiddo*: Harry Pirie-Gordon, "Turkish Campaigns: The Palestine Campaign," *Encyclopaedia Britannica: The New Volumes*, vol. 17, 12th ed. (London: Encyclopaedia Britannica Company, 1921), 824.

marching toward Damascus: Fromkin, *A Peace to End All Peace*, 333.

Bulgaria succumbed: Mick, "1918: Endgame," 160.

signed an armistice: Fromkin, *A Peace to End All Peace*, 363–64.

Danube River: Ibid.

topic of peace: Ibid., 367.

270 *flooding back toward home*: von Sanders, *Five Years in Turkey*, 306.

destroyed their ammunition: Ibid., 307.

seized the city: Pirie-Gordon, "Turkish Campaigns," 824.

Beirut followed, then Aleppo: Ibid., 825.

cabinet of the Young Turks to resign: Taner Akçam, *A Shameful Act: The Armenian Genocide and the Question of Turkish Responsibility* (New York: Metropolitan Books, 2006), 243.

October 30, 1918: Philip Mansel, *Constantinople: City of the World's Desire, 1453–1924* (London: John Murray, 2006), 378.

cease-fire: Robin Prior, "The Ottoman Front," in CHFWW, 318–19.

Hostilities ended the very next day: Charles Townshend, *Desert Hell: The British Invasion of Mesopotamia* (Cambridge, MA: Belknap Press of Harvard University Press, 2011), 434.

terms of their armistice: Mansel, *Constantinople*, 378.

collapse of the Austro-Hungarian Empire: Mick, "1918: Endgame," 166.

Germans signed an armistice: Ibid., 163–64.

271 *ten million had died*: "Chain of Events that Upset World: Review of the War That Cost 10,000,000 Lives and $50,000,000,000 in Property," *New York Times*, November 17, 1918.

twenty-one million had been wounded: Bruno Cabanes, "1919: Aftermath," in CHFWW, 172.

walled with Allied warships: Mansel, *Constantinople*, 379–80.

parliament was dissolved: Akçam, *A Shameful Act*, 245.

former leaders of the CUP vanished: Mansel, *Constantinople*, 378.

"Where have the pashas gone": Akçam, *A Shameful Act*, 245.

272 *An extradition order*: "L'extradition de Talaat," *Renaissance*, December 9, 1918.

close-quartered servicemen: Pirie-Gordon, "Turkish Campaigns," 825.

twenty million dead: Fromkin, *Peace to End All Peace*, 379.

lowered the Ottoman flag: Donald Bloxham, *The Great Game of Genocide: Imperialism, Nationalism, and the Destruction of the Ottoman Armenians* (Oxford: Oxford University Press, 2009), 153.

December 21, 1918: Sina Akşin, "Franco-Turkish Relations at the End of 1919," *Essays in Ottoman-Turkish Political History* (Istanbul: The Isis Press, 2000), 57.

273 *French pushed into Cilicia*: Ibid., 152.
Légion d'Orient: Ibid., 140.
Armenian volunteers: Ibid., 141–43, 151.
survivors of the massacres: Stepan Dardooni, "Seminarian, Deportee, and Legionnaire," in *The Cilician Armenian Ordeal*, ed. Paren Kazanjian (Boston: Hye Intentions, 1989), 207.
revenge on the Turks: Akçam, *A Shameful Act*, 340; Bloxham, *The Great Game of Genocide*, 152.

275 *great fanfare*: Associated Press, "Bugles Greet Delegates: Wilson's Arrival the Signal for a Popular Demonstration," *New York Times*, January 19, 1919.
some two dozen nations: Cabanes, "1919: Aftermath," 175.
League of Nations: Associated Press, "League Structure Becoming Definite," *New York Times*, January 21, 1919.
presided over the negotiations: Cabanes, "1919: Aftermath," 175–76.
the drawing of new borders: Ibid., 176–77.
Armenian mandate: "Anxious to Hand US Armenian Mandate," *New York Times*, March 1, 1919.
how about Palestine: Fromkin, *A Peace to End All Peace*, 397.
Italy wanted its share: "Summary of Allied Aims: Territorial Ambitions Will Be Laid Before Conference," *New York Times*, February 3, 1919.
Arab federation: "Prince of Hedjaz in Paris," *New York Times*, January 20, 1919.
al-Hejaz: "Arabs Look to US for Independence," *New York Times*, January 22, 1919.
Neither was Russia: Cabanes, "1919: Aftermath," 175.
weakened Turks would regroup: Fromkin, *A Peace to End All Peace*, 398.

276 *Piles of ashes*: Kaloustian interviews.
All that was important to their community: "A Ada-Bazar" [In Adabazar], *Renaissance*, December 21, 1918.
house of worship: Hovnatan Geulerian, "Life in Adabazar," *Badmaqirk*, 737–38.
denuded of its gravestones: Ibid., 738.
"a mass of ruins": "A Ada-Bazar."

277 *A telegram*: Turkish Republic Prime Ministry General Directorate of the State Archives, "That Armenians Are Allowed to Return," in *Armenians in Ottoman Documents, 1915–1920* (Ankara: Başbakanlık Devlet Arşivleri Genel Müdürlüğü, 1995), 185.
Some had even taken two: "A Ada-Bazar."
like a scarecrow: Sos-Vani, "Forced Exile," in *Badmaqirk*, 726.
there was nothing: Kaloustian interviews.

278 *black markings striped the window*: Ibid.
neighborhood had been razed: Ibid.
screams of the dying: Ibid.

Epilogue

283 Armenian Genocide Martyrs Monument: "Massive 15-ft. Armenian Genocide Unveiling April 1st," Armenian Genocide Martyrs Monument, http://armenianmonument.org.

284 *reopened schools:* Sos-Vani, "Forced Exile," in *Badmaqirk,* 725.
shelters for the refugees: Ibid., 724–25.
weekly rations: Hovnatan Geulerian, "Life in Adapazar," in *Badmaqirk,* 738.
Three were Tevon's: "Hrachouhi and Krikor Haroutijinian," Gomidas Institute, "Armenian Orphans in the Care of the Armenian Relief Organisation of Istanbul as of October 31, 1919," http://gomidas.org.
a gaping hole in its heart: Geulerian, "Life in Adapazar," 738.
spirit hadn't been broken: Sos-Vani, "Forced Exile," 725.
Archbishop Stepannos Hovagimian had survived: Ibid., 726.

285 *shadow of the missing belfry:* Geulerian, "Life in Adabazar," 737–38.
overseeing the dispersed Greek, Italian, and French troops: "Chronological Review of the World's Remaking," *New York Times,* January 4, 1920.
"against humanity and civilization": This and subsequent information concerning the tribunals is per Gary Jonathan Bass, *Stay the Hand of Vengeance: The Politics of War Crimes Tribunals* (Princeton, NJ: Princeton University Press, 2000), 106, 123–29.
Ibrahim Bey: Vahakn N. Dadrian and Taner Akçam, *Judgment at Istanbul: The Armenian Genocide Trials* (New York: Berghahn Books, 2011), 216–17.

286 *forces loyal to the sultan moved into Adabazar:* Sos-Vani, "The Kemalist Era," in *Badmaqirk,* 728–29.
expecting severe conditions postwar: Arnold J. Toynbee, *The Western Question in Greece and Turkey: A Study in the Contact of Civilisations* (London: Constable and Company, 1922), 187–89.
two hundred thousand people protested: Philip Mansel, *Constantinople: City of the World's Desire, 1453–1924* (London: John Murray, 2006), 389.
sultan's supporters: Sos-Vani, "The Kemalist Era," 728–29.
and the Greeks: SM2, 92; W. J. Childs, "Greeks Drive Turks from Eski-shehr," *New York Times,* March 30, 1921; Edwin L. James, "Kemal Menaces British Forces," *New York Times,* April 2, 1920.
sought refuge in nearby towns: Kaloustian interviews.
Adabazar changed hands: SM2, 92.
other Christians followed: Toynbee, *The Western Question,* 241–42.
all their possessions behind: Kaloustian interviews.
brother-in-law died from an illness: Ibid.
boarded a boat to Constantinople: SM2, 92.
80 percent of the six hundred Armenians: Badmaqirk, 457.
seizing Smyrna: David Fromkin, *A Peace to End All Peace* (New York: Henry Holt, 1989), 545–46.

rushing toward the quay: Ibid.

Tens of thousands more Armenians: Dadrian and Akçam, *Judgment at Istanbul*, xviii.

countless Greeks: RK, 721.

287 *several hundred thousand:* "Posthumous Memoirs of Talaat Pasha," in *The Armenian Genocide: News Accounts from the American Press, 1915–1922*, ed. Richard Diran Kloian (Richmond, CA: Heritage, 2007), 357.

had to be double that: YDZ1, 19.

victory over the Greeks: "Why Turks Refuse to Sign; Demand Real Independence," *New York Times*, March 18, 1923.

Treaty of Lausanne: "Impressive Local Demonstration," *New York Times*, July 25, 1923.

Kemal its first president: "Angora Elects Kemal," *New York Times*, August 14, 1923.

Atatürk: "Atatürk," *New York Times*, January 13, 1935.

memory of the diverse people: Fatma Müge Göçek, "Reading Genocide: Turkish Historiography on 1915," in *A Question of Genocide: Armenians and Turks at the End of the Ottoman Empire*, ed. Ronald Grigor Suny, Fatma Müge Göçek, and Norman M. Naimark (New York: Oxford University Press, 2011), 44–45.

hadn't referred to the killings as genocide: Office of the Press Secretary, "Statement by the President on Armenian Remembrance Day," April 24, 2012, http://www.whitehouse.gov.

lost faith in the Turkish courts: Bass, *Stay the Hand of Vengeance*, 128.

288 *prisoner swap:* Donald Bloxham, *The Great Game of Genocide* (Oxford: Oxford University Press, 2009), 162–64.

While strolling through Berlin: "Talaat Pasha Slain in Berlin Suburb," *New York Times*, March 16, 1921.

Soghomon Tehlirian, had lost his entire family: "Says Mother's Ghost Ordered Him to Kill: Armenian on Trial in Berlin for Murder of Talaat Pasha Reveals Vision," *New York Times*, June 3, 1921.

German jury acquitted him: "Armenian Acquitted for Killing Talaat," *New York Times*, June 4, 1921.

Operation Nemesis: Bass, *Stay the Hand of Vengeance*, 145.

was completely smitten: Anahid MacKeen's memories of her father and mother.

Zaruhi departed for France: Kaloustian interviews.

crooked priest: Anahid MacKeen's recollections of conversations with her father.

289 *"It's too hard for me to be at joyful occasions":* Anahid MacKeen's recollection of her father.

290 *Arab Spring:* Fouad Ajami, "The Arab Spring at One," *Foreign Affairs* (March/April 2012).

293 *two Armenians gunned down:* Robert Lindsey, "Turkish Diplomat Is Slain on Coast," *New York Times*, January 29, 1982.

"I Apologize": Esra Özyürek, "A Turkish 'I Apologize' Campaign to Armenians," *Los Angeles Times*, January 5, 2009.

just before the ninety-ninth anniversary: "Erdoğan Offers Turkey's First Condolences to Armenians for 1915," *Today's Zaman,* April 24, 2014.

"inhumane consequences": "Turkish Prime Minister Mr. Recep Tayyip Erdoğan Published a Message on the Events of 1915," Republic of Turkey, Ministry of Foreign Affairs, http://www.mfa.gov.tr.